THE CONSTELLATION

THE CONSTELLATION
Lockheed's Graceful Masterpiece

Alexander Clifton

SunRise

First published in Great Britain in 2021 by SunRise

SunRise Publishing Ltd
Kemp House
152-160 City Road
London EC1V 2NX

ISBN 978-1-9162161-4-3

A CIP catalogue record for this book is available from the British Library.

Typeset in Bookman Old Style, titles in Impact, captions in Acumin.

Printed by Kingsbury Press, Doncaster, England.

Contents

Introduction

The airline business is a hundred years old. In that time, uncountable airliners have been conceived, designed and built but, for all their diversity, less than a hundred types have ever sold in large numbers and, unlike military aircraft, only a handful are truly iconic. The shortlist, in fact, is so brief (rarely more than six), they can be named in a sentence. Because such rankings can spark passionate debates among the cognoscenti, I shall leave you to write your own. All I will say is that you will be hard-pressed to leave out one of the few commercial aircraft that still stirs hearts across the world: Lockheed's graceful masterpiece — the Constellation. Her elliptical wings, triple tail, insect legs and dolphin-shaped fuselage still make her instantly recognisable to almost everyone, even those who could identify few other aeroplanes. For post-war travellers she came to symbolise panache and elegance in what is sometimes known as flying's 'Golden Age'. Today, eighty years after her birth, she is loved by another generation for her retro style and 1940s glamour.

My own love affair with the 'Connie' began more than fifty years ago on a cold morning at Frankfurt Airport. In the mid-1960s I was travelling to Africa with my family aboard a sparkling new VC10 of BOAC. In those days, BOAC flights often called at Frankfurt, Rome or Zürich to pick up onward passengers. A bus took us to the transit lounge while our aircraft was being refuelled, and we drove past a Super Constellation of Condor Flugdienst, the holiday charter arm of Lufthansa. At the exact moment we were alongside, one of her huge radial engines came to life with a billow of smoke and flame, closely followed by another. In my mind's eye, I can still see her pale blue and yellow livery turning opaque in the smoke that enveloped our bus, along with the faces of curious passengers behind her windows. I also remember the smell of high-octane fuel and the muffled roar of her engines as they fired up. I knew that these magnificent aircraft would soon be made obsolete by the new jets then coming into service, and I felt a pang of regret.

Decades later, on a business trip to Karachi, I stayed overnight at what is now the Airport Hotel but had been, until 1973, the Speedbird Rest House. Built by BOAC in 1955, it originally provided somewhat austere accommodation for BOAC and Qantas crews. While taking a stroll around the vicinity, I saw the unmistakable profile of a Connie. The remains of a Pakistan International Airlines (PIA) Super Constellation had been parked on a strip of land not far from the aiport perimeter. Her PIA titles were faded, but still visible, her windows had been crazed by the sun and many removable parts had been plundered from her, but she still managed to look majestic. There were steps up to her open forward door and I mounted them — it was the first time I had ever been onboard a Connie. Most of her seats and original cabin interior had survived and, for a few moments, I could imagine what it must have been like to fly on her. The pang for this aeroplane that I had felt as a child in Frankfurt returned. It has never left me; I feel it still, all these years later.

1 Howard Hughes and the Birth of a Legend

New airliners, with few exceptions, are created by engineers under the direction of airlines, governments and the boards of aircraft manufacturers. The Constellation, however, was fathered by an eccentric billionaire whose previous contributions to aerospace history had been a series of record-breaking flights and a movie, *Hell's Angels* (1930). As Robert Serling put it: 'From the era of the DC-3 to the dawn of the jet age, Howard Hughes dominated TWA, shaping its destiny as one would mould a piece of clay'. [1]

Hughes wasn't the founder of Trans World Airlines (TWA) but by 1940 he controlled three-quarters of the shares. He was a brilliant record-breaking pilot who had inherited an immense fortune from his father. His first loves were for science, radio and the cinema, but Hughes later became fascinated, if not obsessed, by every detail of the airline business, even to the extent that he'd assumed a false name (Charles Howard) so that he could work for American Airlines anonymously as a co-pilot. He had first dealt with TWA in 1936, when he bought the original and only DC-1 from them to attempt a record-breaking flight around the world. TWA had withdrawn the aircraft from passenger service and used it as a flying laboratory for high altitude research. The engines had been changed to Wright GR 1820 F55 Cyclones with two speed blowers, and Hamilton Standard constant speed propellers.

1 Serling, Robert J. Howard. *Hughes' Airline: An Informal History of TWA*. Lume Books, 1983.

Hughes later switched to a faster Lockheed L-14 and is believed to have forgotten where he left the DC-1, until he was reminded by a telephone call from Burbank Airport enquiring how long he planned to leave it there!

Before the Constellation, Howard Hughes (below and opposite) was known as an industrialist, film producer and record-breaking pilot.

To Al
Best of luck
Howard

TWA had been the second customer (after Pan Am) for the Boeing 307 Stratoliner, a civilian development of their B-17 bomber. The Stratoliner was the world's first pressurised airliner; its 20,000-foot ceiling and 1,750-mile range enabled it to cruise for long distances above the weather, but its entry into service was plagued by accidents. The first aircraft crashed during a demonstration flight for KLM, killing all the occupants including TWA's chief pilot. The first Stratoliner delivered to TWA suffered a multiple engine failure and made a forced landing in an open field (mercifully, without injuries), during a proving flight from Kansas City to Albuquerque. These and other accidents did nothing to enhance the plane's reputation and the two biggest airlines, United and American, placed no orders, preferring to wait for the Douglas DC-4. Douglas was then the most trusted company for airliners, and by 1939 their DC-3 was carrying a staggering 90 per cent of the world's commercial passengers.

The Boeing 307 Stratoliner was the world's first pressurised airliner, but few were sold (Lockon Aviation Photography).

In the late 1930s Douglas was the 'go to' manufacturer for airliners. Pictured is the DC-4E featuring a triple tail.

TWA inaugurated Stratoliner service on 8 July, 1940, with a flight from La Guardia to Burbank carrying the movie stars Tyrone Power and Paulette Goddard. The flight time was fourteen hours and nine minutes with stops in Chicago, Kansas City and Albuquerque — nearly four hours quicker than the DC-3s used by most of TWA's competitors. Although the new aircraft quickly became popular with passengers, only ten Boeing 307s had been completed by the time the United States entered the Second World War, one of which (NX-19904), was sold to Howard Hughes for another attempt at a round-the-world flight. Hughes hoped to break his own record of 91 hours 14 minutes set from July 10–14, 1938 in his Lockheed 14. This machine was fitted with extra fuel tanks and was ready to set out on the first leg when Germany invaded Poland (1 September, 1939), putting an end to Hughes' plans. After the Second World War, TWA sold all five of their Stratoliners to the French airline, Aigle Azure, who continued to operate them until the late 1960s.

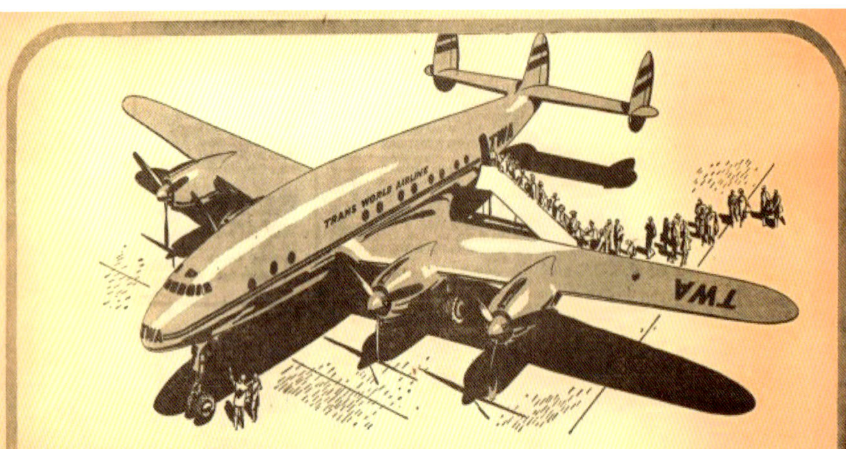

★ TWA WILL BE FIRST TO FLY CONSTELLATIONS ACROSS THE ATLANTIC

Soon TWA will take delivery of the first commercial Constellations to come out of the great Lockheed plant — the first of a new 30 million dollar fleet.

Before long, TWA will begin the first Constellation service across the Atlantic, and soon half way round the world to India.

The Constellation, developed orginally for TWA, has been released by TWA for sale to other airlines. Eventually it will be flown by many of the leading airlines of the world.

★ TRANSATLANTIC FLYING TIME WILL BE CUT

Today it requires about a day to fly across the Atlantic. Using Constellations, TWA will cut this time nearly ten hours. Note the illustrations of time-savings in the next column —

New York To	Present Time	TWA Time	Time Saved
Foynes, Eire	23 hrs. 25 mins.	13 hrs. 50 mins.	9 hrs. 35 mins.
Lisbon, Portugal	29 hrs. 30 mins.	18 hrs. 10 mins.	11 hrs. 20 mins.

★ TRANSATLANTIC FARES WILL BE CUT TO REASONABLE LEVELS

The air fare from the United States to Foynes, Eire during 1939 was $337.00. After the war broke out, this fare was raised to $625.00. Today it is $525.00.

This North-Atlantic fare of over 17¢ per mile is *nearly four times the rate charged by competing airlines for travel in the United States.* TWA will establish reasonable fares for transatlantic traffic.

★ TRANSATLANTIC SERVICE WILL HAVE THE SAFETY AND COMFORT OF OVER-WEATHER FLYING

The super-charged cabins of TWA's Constellations are air conditioned for high-altitude flying. This means that you can fly over the bad weather in the smooth upper air for greater safety and comfort.

Service on TWA's Trans World Airline will begin soon.

Schedules will be announced and supplied to your travel agent shortly.

POINTS THE WAY

Fly the Finest...
906
N6906C
TWA
FLY TWA
TRANS WORLD AIRLINES

Tyrone Power and Paulette Goddard arrive at Burbank on TWA's inaugural Stratoliner service in 1940.

For all its popularity, the Boeing 307, which carried 33 passengers by day and 16 in sleeping berths at night, was too small to be operated profitably by TWA, and Hughes began to dream of a bigger aircraft. Much has been written about his flair for business coupled with a chronic inability to make decisions, and a level of social dysfunctionality that would eventually turn him into a recluse. Among his many gifts, however, was that crucial talent which is sometimes granted to a genius: insatiable curiosity coupled with a refusal to accept no for an answer. Once Howard Hughes had made up his mind that he wanted a bigger and faster plane, he wasn't going to be refused by anyone. In the preceding years he had commissioned and purchased aircraft from each of the three leading manufacturers of the day: Boeing, Lockheed and Douglas. None had found him an easy customer to satisfy, but Hughes had a particular respect for Hall Hibbard, Lockheed's chief engineer, and his young assistant, Clarence 'Kelly' Johnson. This was the team that had designed the Lockheed 14 in which Hughes had claimed the record for a round-the-world flight.

Howard Hughes with the Boeing 307 Stratoliner.

SUPER-
SUPER-
N6937C
SUPER-
TWA

WORLD AIRLINES
TWA
TWA

2 Lockheed

The Lockheed Corporation had a history almost as eccentric as Hughes' own. Their founder, Allan Loughead, was raised in Santa Barbara, California by his mother, who later moved to a fruit ranch where he and his brothers experimented with gliders. In 1909, Allan's half-brother, Victor, wrote a book: *Vehicles of the Air,* which became influential in aircraft design and in the same year Allan began to race cars. The brothers were soon building and converting aircraft and gliders for

The Loughead brothers Malcolm (left) and Allan at the controls of their F-1 flying boat.

a Chicago automobile dealer, James E Plew. When two of Plew's professional pilots were unable to get a machine airborne, Allan boasted: 'I've got a $20 gold piece that says I'll make it fly, and I'm offering three-to-one odds! Any takers?' There were no takers, but he got the airplane airborne on his second try. As he later said, 'It was partly nerve, partly confidence and partly damn foolishness. But now I was an aviator!'

By 1912, Allan and his brother Malcolm were building seaplanes and taking passengers on sightseeing flights across California. By 1916, the brothers had based the Loughead Aircraft Manufacturing Company in Santa Barbara and were building their 10-place, twin-engined F-1 flying boat (huge by the standards of the day) with an eye on the aerial sightseeing business. In 1917, when the United States entered the First World War, the company got a contract to build two Curtiss flying boats plus an agreement from the Navy to test the F-1. Allan Loughead, with a crew of three, flew the prototype from Santa Barbara to San Diego in April 1918. They broke the record convincingly but failed to win either a Navy or Army contract. By November, the end of the First World War had dashed all hopes of future military orders and in 1920 the company folded. Allan went into the real estate business and Malcolm sold brake systems for cars.

Six years would pass before Allan Loughead, along with John Northrop, Kenneth Kay and Fred Keeler, had raised enough money to form the Lockheed Aircraft Company in Hollywood. The company name was now spelled phonetically to avoid mispronunciations and, possibly, to distance themselves from the earlier failure. The 'Roaring Twenties' was a period of sharp economic growth in America and the new company found a ready market for their Vega. The rugged construction and long range of this high-wing, six-passenger monoplane airliner helped it to find fame in the hands of a series of legendary record-breaking pilots including Amelia Earhart, Wiley Post, and George Hubert Wilkins. In 1928, with annual sales exceeding one million dollars, Lockheed would relocate to Burbank, California, where they would remain for more than 50 years. By 1929, 300 workers were building five aircraft per week but, in July, the majority shareholder, Fred Keeler, sold 87% of the Lockheed Aircraft Company to Detroit Aircraft Corporation. Allan Loughead resigned

Amelia Earhart's Lockheed Vega 5B.

In 1930, Wiley Post gained international fame by flying his Lockheed Vega, *Winnie May*, around the world. Fifty thousand people greeted him on his return after a flight of 7 days, 18 hours, 49 minutes. The aicraft is now preserved in the National Air and Space Museum.

a month later. The Great Depression was biting, and it would soon have a devastating effect on aircraft sales, just as the Great War had done a decade before.

Above: the Lockheed Aircraft factory in Burbank, California, prior to the Second World War. Below: the same factory after camouflage had been applied during the war (US Army Corps of Engineers).

Detroit Aircraft went bankrupt and remained so until 1932 when the brothers Robert and Courtlandt Gross, along with Walter Varney, bought the company out of receivership for a mere $40,000. Ironically, Allan Loughead had raised $50,000 with the same goal in mind, but chose not to bid, thinking he had too little capital to succeed. The new company was called the Lockheed Aircraft Corporation and it would continue under

that name until 1977, when it became the Lockheed Corporation. The revived Lockheed continued to build the Vega, and spent $139,400 developing the Model 10 Electra, an advanced twin-engine transport of which they sold 40 in the first year of production. In 1937, Amelia Earhart and her navigator, Fred Noonan, would tragically disappear over the Pacific while attempting to fly their Electra around the world. Neither the aircraft, nor their remains, were ever found, and speculation about their fate continues to this day. The disaster was to become one of the most infamous aircraft mysteries in history but that did not prevent the subsequent Lockheed Model 12 Electra Junior and the Lockheed Model 14 Super Electra from widening Lockheed's reputation for innovative high-performance aircraft. One of them, the Lockheed Model 14, was developed into the Hudson bomber, which sold in significant numbers to the Royal Air Force (RAF) and United States Army Air Force (USAAF) before and during the Second World War.

Amelia Earhart and Fred Noonan board their Lockheed Electra during a round-the-world attempt which would end in tragedy and cost them their lives.

The Lockheed 14 was developed into the Hudson light bomber and coastal reconaissance aircraft. Nearly 3,000 were built and operated by various Allied air forces during the Second World War and by commercial operators afterwards.

Lockheed's Kelly Johnson, who would go on to lead the legendary 'Skunk' works and design some of the world's most celebrated aircraft (these include the U-2 and SR-71 Blackbird spy planes), joined Lockheed in 1933 as a tool designer. His boss, Hall Hibbard, would later remark: 'That damned Swede can actually see air.' One of the first Lockheed aircraft to bear Hibbard and Johnson's joint signatures was the P-38 Lightning fighter. Like many of Johnson's later designs, it would depart from the orthodoxy of the day. Among other innovations, it featured a twin boom tail while housing the pilot in a central nacelle. It was the first military airplane to exceed 400 miles per hour in level flight — so fast, in fact, that the Lightning also became the first aircraft to experience compressibility problems which, at the time, were little understood.

The legendary Kelly Johnson (below and opposite) would later lead Lockheed's 'Skunk' works which designed many celebrated aircraft including the U-2 and SR-71 Blackbird spy planes. He is pictured in and beside a Lockheed F-104 Starfighter.

RESCUE

3 Jack Frye and TWA

If Howard Hughes and Kelly Johnson were the parents of the Lockheed Constellation, then Jack Frye was the godfather. Although less has been written about him than other airliner pioneers of the period, his influence on the nascent business was, in truth, immense. As one author noted, 'Jack Frye of TWA may very well have been the most underrated and unappreciated airline president of them all. He was a pilot himself, smart and as likeable as he was capable, but was also saddled with the fact that he was overshadowed and subservient to TWA's majority stockholder, who controlled TWA and happened to be Howard Hughes'.[1]

TWA's likeable Jack Frye may have been the most underrated airline president of them all.

1 Serling, Robert J, *The President's Plane Is Missing*, Cassel & Co 1968.

TWA owned the original and only Douglas DC-1; they later sold it to Howard Hughes.

As early as 1926 Frye and his partners had formed Standard Airlines which began operations with a single Fokker F-7 aircraft flying from Los Angeles to Tucson with a stop at Phoenix. By 1930 Standard had been acquired by Western Air Express which merged with Transcontinental Air Transport to form Transcontinental and Western Air, the airline which eventually would become TWA. In 1932, Frye's search for a better aircraft had persuaded Douglas to produce the DC-1, followed by the DC-2 and DC-3. TWA took delivery of the only DC-1, and Frye, along with Eddie Rickenbacker, then president of Eastern Air Lines, piloted the machine to a new transcontinental record of 13 hours and 4 minutes. Frye became president in 1934 at a time when John Herz, the yellow cab and car rental mogul, was the majority shareholder. When Herz recoiled at the cost of TWA's order for Boeing 307s, Frye and Rickenbacker turned to Howard Hughes and persuaded him, to practically everyone's surprise, to take a controlling interest in the company. Months before

517
N90831
TWA
TWA
G207

ANS WORLD AIRLINE
13

the 307s had even been delivered, Hughes was already contemplating an aircraft which could fly from coast to coast nonstop.

By 1939 Howard Hughes had challenged Lockheed to build an airliner that could fly at up to 300 miles-an-hour with the range to go from west coast to east coast in eight hours nonstop, and westbound in nine hours with one stop. The company was already working on their unpressurised Excalibur, a four-engine design for Pan American for which they had built a wooden mock-up. This unlikely aircraft featured, on the insistence of Pan Am, a wing thick enough to allow an engineer to reach the engines in flight. Hughes rejected Excalibur out of hand, demanding a 40-passenger aircraft with sleeping berths for twenty, a 3,500-mile range and a payload of 6,000 pounds. Lockheed's response was to suggest Wright's new and relatively untried R-3350 power plant, then the world's most powerful aircraft engine. This twin-row, supercharged, radial engine, comprising 18 cylinders and displacing some 55 litres, could produce more than 2,000 horsepower. With four of these, Lockheed believed, they could 'up the ante' considerably and build a machine capable of carrying sixty passengers by day at a cost of $450,000 per aircraft, but, as is often the case with ambitious all-new planes, when finally delivered the aircraft cost significantly more. Hall Hibbard would later recall, 'Up to that time we were sort of "small-time guys", but when we got to the Constellation we had to be "big time guys...". We had to be right and we had to be good'.

4 Early Designs

An early and crucial decision was to use a scaled-up version of the P-38 Lightning's elliptical wing, which was known to give good lift and low drag at high altitude, along with a triple tail whose low profile would allow maintenance within typically sized hangars of the day. The fuselage was a greater challenge — while a tube takes pressurisation stresses well and is cheaper to build, it is aerodynamically less efficient. To meet Hughes' demands for performance, Kelly Johnson decided to go with a dolphin-shaped fuselage in which no two bulkheads would be the same size; the fuselage was slightly turned up at the rear, to keep the triple tail out of the propeller wash, and slightly turned down at the front to avoid the nose gear being impossibly long. This also permitted him to increase the cabin roof height over the wing spar with a stepped-up floor, creating more passenger seating without any increase in drag. It was a difficult shape to design and build, but the result was a more efficient aircraft. Because of the enormous diameter of the propellers (16' 10"), the landing gears, especially the nosewheel, were very long. Tricycle landing gear, which Lockheed had used with their P-38, was then an innovation and had only appeared on one other airliner, the DC-4. With a wingspan of 123 feet and a length of 95 feet the Constellation was a very large aircraft for her time. The distinctive outlines of the wing, fuselage and triple tail combined with her long, insect-like landing gears, would become unmistakable components of the Connie's unique 'look'. Johnson had created a triumph of form over function, an efficient and fine-looking aircraft which is still celebrated as one of the most beautiful ever.

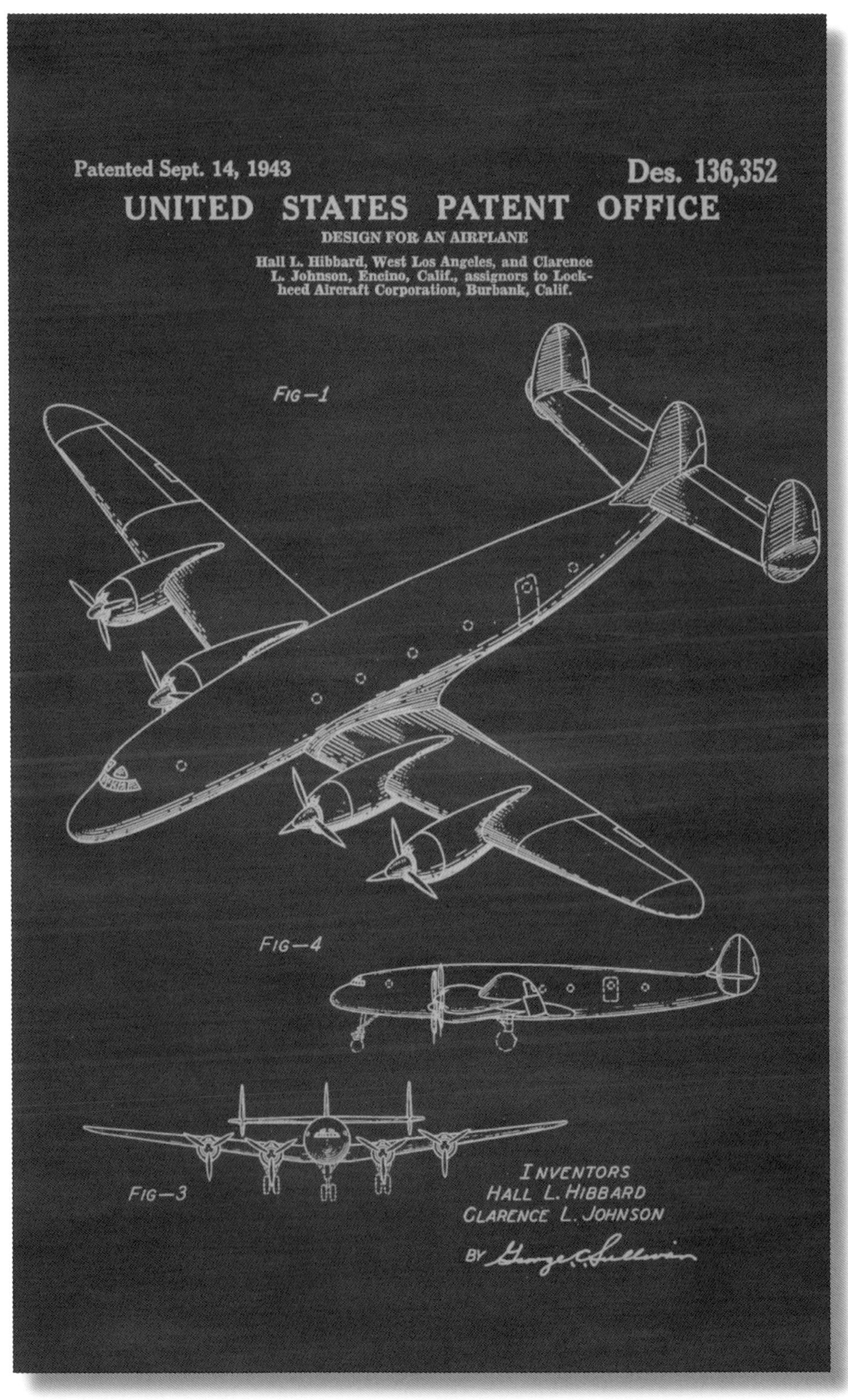
Patented Sept. 14, 1943
Des. 136,352
UNITED STATES PATENT OFFICE
DESIGN FOR AN AIRPLANE
Hall L. Hibbard, West Los Angeles, and Clarence L. Johnson, Encino, Calif., assignors to Lockheed Aircraft Corporation, Burbank, Calif.
FIG—1
FIG—4
FIG—3
INVENTORS
HALL L. HIBBARD
CLARENCE L. JOHNSON
BY

Bob Gross, who was Lockheed's president from 1934 to 1956, led the company throughout the design, construction and entry into service of the Constellation.

Innovation, however, was not confined to her shape. Today, airlines are reluctant to buy aircraft which incorporate too much new technology in a single step; they prefer either a proven engine in a new airframe, or a new engine in a proven airframe. Lockheed, however, knew no such inhibitions and the Connie was crammed with new ideas from nose to tail. She was the first commercial aircraft to have hydraulically boosted controls; without them, crews would have encountered huge control loadings because of her size and speed. Pilots were consulted as to the best boost ratio: they chose 9.33:1 for the elevators (i.e., the pilot's effort boosted 9.33 times), 23:1 for the rudders and 26:1 for the ailerons.

The design of the cockpit set new standards for the industry and became the blueprint for future aircraft. When de Havilland was designing the Comet jet airliner, they adopted the Constellation cockpit layout because so many pilots were already used to it. The Connie's nine cockpit windows were angled for minimum drag, which meant the pilots had to sit quite close to them for good visibility and that, in turn, led to a tightly spaced flight deck. As well as dual controls, the flight instruments were duplicated on both sides. The flight engineer sat behind the co-pilot, facing outward, and operated most of the engine controls. Behind the engineer, and facing backwards, sat the radio operator. Further back, behind the cockpit bulkhead, was a crew rest area with an optional position for a navigator.

The enormous diameter of the Constellation's propellers meant that she needed exceptionally long landing gears (Smithsonian Air and Space Museum).

The cockpit of the Constellation set new standards and became the blueprint for many airliners which followed (Russell E Lown).

This was an era in which most airlines operated on an all-first-class basis; high-density economy seating would come later. The first Constellations would have twelve rows of five seats with a generous 38-inch seating pitch, while the sleeper version offered 22 berths. When the design was finalised, Hughes responded to Lockheed's Bob Gross by saying, 'TWA can't pay for them. The damned airline's broke. We can't take it to the banks. They'd think we were nuts talking about a three hundred-mile-an-hour airplane. Hell, I guess I'll have to pay for them myself. Go ahead and build 'em, Bob. Send the bills to the Hughes Tool Company'.[1] In a later conversation with TWA's president, Jack Frye, Hughes confided, 'When we get through with 'em, Jack, we'll have one hell of an airplane!'

Hughes' obsession was aircraft, he was always pressing manufacturers to build faster, bigger, better machines.

---

1 Serling, Robert J., *Howard Hughes' Airline: An Informal History of TWA*, Lume Books, 1983

Frye, on the other hand, saw the bigger picture and understood that aircraft could only make money within a well-structured, well-managed, ambitious airline that had the right routes to fly them on. While Hughes could take much of the credit for the Constellation, Frye could take the credit for building TWA from a small regional carrier into an international giant that could challenge the likes of Pan American, United and American.

SUID-AFRIKAANSE-LU

5 First Orders

Early in 1940 Hughes signed an order for nine aircraft, which was later increased to forty in exchange for which Lockheed guaranteed him the first thirty-five machines off the assembly line. At $18 million, it was then the biggest single order in the history of the commercial aircraft business. Hughes believed the new plane would allow him to outpace United and American, who were waiting for the Douglas DC-4. So closely did he value secrecy, that he wouldn't allow anyone from Lockheed to type the contract, asking the wife of one of the TWA managers, who was a court stenographer, to prepare it for him. In a possible harbinger of the mental health problems that would later dog him, he insisted on all the TWA staff who knew of the project using only code names: Jack Frye was Jesus Christ and Hughes himself was God!

Lockheed designated the project the 0-49, but the aircraft would later be dubbed the Constellation or, more affectionately, the 'Connie'.[1] The design was developed in such strict secrecy that the handful of TWA employees who knew about it couldn't even tell their wives or families, never mind their colleagues, what they were doing, meaning that their personal relationships often deteriorated to the point of animosity. Hughes paid enormous attention to detail and, when the project reached the stage where Lockheed had built a full-size wooden mock-up, Hughes would sit in it and constantly make minute changes, especially to the décor and colour schemes. Kelly Johnson had designed the airframe, but Hughes' fingerprints were all over the rest of the machine.

The Wright R-3350 Turbo Compound Radial Engine which powered the 0-49 Constellation.

1 Eastern Air Lines' president, Eddie Rickenbacker, forbade his crews to use the nickname 'Connie', believing it to be effeminate.

While TWA waited for the Constellation, United and American had pinned their hopes on the Douglas DC-4.

Once Lockheed were able to publicly announce the new plane, in 1941, it caused a media sensation and they quickly got an order from Pan American for a further 40 (ten 049s and 30 of the longer-range 349s), while KLM, the Dutch airline, ordered ten. History, however, was about to play her hand in a decisive way. By 1941 the United States government was already contemplating involvement in the Second World War, which had been raging in Europe since 1939. They set up a Commercial Aircraft Priority Committee and took control of the distribution of all commercial aircraft. Before long, the Army was drafting civil airliners directly from the production line. They quickly saw the potential of the Constellation as a fast, long-range transport aircraft and placed an order for 180, designated C-69s. The Army would own the planes, but the airlines would operate some of them under contract to the Army. The Constellation would spend the duration of the war in military camouflage and the public would have to wait.

Howard Hughes personally designed much of the Constellation's interior, paying particular attention to fabrics and colours.
Above: first class cabin of a BOAC Constellation.
Below: economy class cabin of a KLM Constellation.

There would, however, be just one exception: it was agreed that the very first production Constellation off the assembly line would be symbolically delivered to the airline, TWA, before being handed straight back to the military.

6 The Connie Flies

The first prototype Constellation (civil registration NX25600) flew on January 9, 1943, on a short ferry hop from Burbank to Muroc Field (today's Edwards Air Force Base) for testing. The Army insisted on Edmund T. 'Eddie' Allen (on loan from Boeing) being the pilot in command, while Lockheed's own Chief Test Pilot, Milo Burcham, had to suffer the indignity of being the mere co-pilot. Kelly Johnson and his assistant, Rudy Thoren, were also on board along with Chief Mechanic Dick Stanton. When the 58-minute flight ended, Allen remarked, 'This machine works so well that you don't need me anymore!' That first flight had, indeed gone well, so well, in fact, that five more flights were performed

TOTAL

on the same day. The following morning the *Los Angeles Times* enthused at length about the new plane:

SUPER TRANSPORT PLANE IN DEBUT

Lockheed's Air Marvel Makes First Flight; Believed to Be World's Largest and Fastest; Built Like Fighter, Can Outspeed Jap Zero

BY MARVIN MILES

Into the winter sky yesterday swept a brilliant new star—Lockheed super-transport Constellation.

First of a galaxy to come, the four-engine colossus sped down the long east-west runway at Lockheed Air Terminal, skipped nimbly off the concrete and boomed upward with the surging roar of 8,000 unleashed horses.

A few breath-taking seconds' full throttle had written a matter-of-fact climax to two years of secret development that evolved a 60-passenger transport faster than a Jap Zero fighter.

There were no fanfares, no speeches—simply an unvarnished war production take-off, emphasizing as nothing else could the grim driving need for huge work planes to carry the battle swiftly to the ends of the earth.

Yet it was the first significant aviation event of 1943. Built along the slim, graceful lines of a fighter the craft is faster than any four-engine bomber now in service. It can cross the continent in less than 9 hours, fly to Honolulu in 12. Even at half power its cruising speed is approximately 100 miles per hour faster than that of a standard airliner!

Within its supercharged cabin, air-density will remain at the 8,000-foot level when the Constellation is cruising at "over-the-weather" altitudes up to 35,000 feet. So great is its power that the monster can maintain 25,000 feet on three engines, 16,500 on two.

As for economy of operation, the new sky queen can fly her full load hour after hour using but one gallon of gasoline per mile.

ONE TAXI TEST

At the controls when the super-transport lifted its tricycle gear in flight were Eddie Allen, Army pilot and veteran four-engine flyer, and Milo Burcham, Lockheed test pilot noted for his substratosphere testing of the P-38. Also, in the ship were C.L. (Kelly) Johnson, chief research engineer for the aircraft company; Rudy Thoren, Johnson's assistant, and Dick Stanton, chief mechanic.

There was but one taxi test yesterday, highlighted by a brief blaze in one of the four engines following a backfire as the ship turned to roll back to the head of the runway.

The fire was doused quickly, and the Constellation stood ready for her maiden flight, her nose into a gentle breeze, the focal point of hundreds of eyes of workers, Army guards and officials watched expectantly. Each engine "revved up" in turn, sending deep-throated echoes over the sun-drenched terminal.

Then the four black propellers whirled as one The Constellation shot forward, the wind in her teeth, a hurtling, bellowing land monster—until her propellers plucked her from the earth in an incredibly short span of runway and sent her thundering triumphantly toward the sun.

GLIDES BACK EASILY

In a moment she had almost vanished, only to bank in a wide turn and drone back over the terminal twice before leading her covey of lesser following craft off toward the desert to the Army airport at Muroc Dry Lake where she gracefully landed an hour later. Shortly before dusk the giant craft returned to the Burbank terminal, slipped down the long "landing groove" of air and settled easily to the runway.

Her debut was over.

Today she will begin the exhaustive test flights to determine her performance before she is turned over to T.W.A. and the Army for the grueling business of war...

General 'Hap' Arnold with Howard Hughes (Colourised by Benoît Vienne).

Howard Hughes boards the prototype Constellation while Jack Frye (with cigar) looks down from the aircraft door (Colourised by Benoît Vienne).

The Army insisted that Boeing's Chief Test Pilot, Eddie Allen (left), should command the first flight of the Constellation while Lockheed's Chief Experimental Test Pilot, Milo Burcham (below right with Kelly Johnson), was the co-pilot. Within a matter of months both pilots would die in flying accidents.

7 The Competition

To fully grasp what a gigantic step forward the Connie had made, we must take a brief look at her only serious competitor, the Douglas DC-4. The overarching success of the DC-3 had established Douglas as the go-to company for airliners, and the four-engined DC-4 was a natural progression which quickly won orders. The prototype first flew in 1942, but, like the Connie, the production of civil versions was delayed by the war. Douglas, nevertheless, built hundreds of the military version, the C-54, and, like the DC-3, she quickly gained a reputation for ruggedness and reliability (a handful still fly commercially to this day). However, although she was the same length and had the same number of engines, she was completely outclassed by later developments of

The C-54 proved to be rugged and reliable and was used extensively during the Berlin Airlift.

SUPER-
G
N6937C
SUPER-
G
TRANS
TWA
TWA

AIRLINES
TWA

the Connie, which flew (in round figures) 50% faster, 60% further and two and a half times higher while carrying nearly three times the payload!

The government kept their promise and allowed the first aircraft off the production line to be symbolically delivered to TWA before being immediately handed over to the USAAF. The acceptance flight, which was from Burbank to Washington, DC, broke records. On April 17, 1944, Hughes and Frye flew the cross-country trip in under seven hours. The Connie averaged 331 mph, flying nonstop in six hours, 57 minutes, and 51 seconds. The actress Ava Gardner, who was Hughes' girlfriend at the time, was intended to be a passenger along with Kelly Johnson but she changed her mind about the flight at the last minute, much to Hughes' annoyance. Johnson had a low opinion of Hughes' piloting skills. 'He damn near killed us', he later claimed.

YLINER April, 1944

Army Men and Statesmen Inspect Ship

Among the persons who congratulated Jack Frye upon the arrival of the Constellation at the Washington National Airport were, left to right, Col. Frank H. Collins, Commanding Officer, ATC, Washington National Airport; Oswald Ryan, (Jack Frye) and Josh Lee. Ryan and Lee are members of the Civil Aeronautics Board.

The Constellation acceptance flight from Burbank to Washington was delayed when Hughes' current girlfriend, Ava Gardner, refused to board the plane. Kelly Johnson later claimed that Hughes 'damn near killed us!' On arrival, the Army were furious to see that Hughes had repainted their airplane in TWA's colours.

The Constellation, new queen of land planes, zoomed across the girth of the nation this month in six hours, 58 minutes, setting a new transport speed record on her maiden flight from Burbank, California to Washington, D. C. Jack Frye, left, and Howard Hughes, right, co-captains of the trip are shown immediately after their arrival in Washington. Jesse Jones, center, Secretary of Commerce, was the first to greet the two pilots of the record flight.

The prototype Constellation, NX25600 with (below) a Lockheed Vega. Unusually for an experimental military aircraft she carried the Lockheed logo on her nose and tail.

Above: NX25600 takes off from Burbank (now Bob Hope) Airport. Below: during test flying at Wright Field, Ohio, Orville Wright boarded the prototype Constellation to make his last ever flight, serving as co-pilot during some trials.

On meeting the Constellation in Washington, the Army were furious to see that Hughes had audaciously repainted the aircraft in TWA's red livery and were little mollified by the fact that it was done with easily removable water-based paint. That night, Hughes held a glittering Washington reception for more than 2,000 VIP guests to make sure everyone knew the Connie had arrived. *The New York Times* enthused that it was, '...an outline of the shape of things to come in air transportation.'

The aircraft was supposed to be handed straight back to the military, but Hughes managed to hold on to it for a few days and did several demonstration flights. On one, with the entire Civil Aeronautics Board (CAB) and a host of other VIPs aboard, he decided to shut down two engines without warning anyone. When told by a TWA vice-president that the passengers were getting rather scared, Hughes responded, 'Hell, I'm gonna cut a third engine, that'll show them what this plane can do!' [1]

When the aircraft was finally returned to the military for service testing at Wright Field, Ohio, Orville Wright, who had made the world's first powered flight, was persuaded by Hughes to fly for the last time (he died in 1948), serving as co-pilot on a test run.

The prototype stayed with the military, and was eventually bought back by Lockheed for use as a test aircraft, but it was never returned to TWA. Despite the publicity, the three-year lead which the Constellation would have given TWA over their competitors was lost. Hughes' only consolation was that he got buy-back rights from the government, which meant that, at the war's end, a sizeable fleet could go into airline service. It would be a long wait, however, before Hughes and Frye finally got their Connies back.

---

1 Serling, Robert J., *Howard Hughes' Airline: An Informal History of TWA,* Lume Books, 1983.

8 The Connie Goes to War

The military version, which would be used as a high-speed, long-range troop transport, was christened C-69. However, although the government had originally placed orders for a total of 202 machines, only 22 had been completed by the end of hostilities, and a paltry thirteen finally went into military service. The reason was a simple one: the government had changed their priorities and now wanted Lockheed to concentrate virtually all their resources into building thousands of P-38 Lightning fighters and Hudson bombers. A staggering 10,037 P-38s and 2,941 Hudsons were completed between 1940 and 1945, but orders for

During the Second World War Lockheed built more than 10,000 examples of their P-38 Lightning fighter.

Southern Cross
847

the remaining Constellations were scaled back and finally cancelled in 1945. Another factor was, in all probability, the Connie's temperamental Wright engines, which caught fire too often for the military's liking. They turned to the more dependable Douglas C-54 Skymaster as an airlifter and told Wright to concentrate on optimising their R-3350 engine for the upcoming Boeing B-29 bomber.

The glamorous Constellation — groomed for military stardom — had been side-lined by a homely workhorse which broke few records and won no beauty contests, but got the job done. The Second World War had meant that Lockheed, along with most of America's other plane makers, had enjoyed years of bulging order books. Early in the war, the British and French governments had placed large orders for American aircraft.

In May 1940, Franklin D Roosevelt called for America to become the 'Arsenal of Democracy' by producing 185,000 aeroplanes,120,000 tanks, 55,000 anti-aircraft guns and 18 million tons of merchant shipping — all to be built in just two years. Hitler had been told by his advisors that this was mere propaganda, but, by the end of the war,

The sole surviving Lockheed C-69 is preserved in TWA livery at the Pima Air & Space Museum (Karsten Palt).

The only Lockheed C-69C was civilianised and delivered to BOAC as G-AKCE.

American companies had, in fact, built nearly twice that number of planes! The awesome manufacturing levels of American industry played a key role in the Allied victory. German production figures for military material never came close and had slowed to a trickle by the final months of the war.

9 The War Ends

When the Second World War ended, Lockheed were a financially strong company. They took a bold decision to keep the C-69 production line open and convert all the surplus and incomplete C-69s to civilian specifications. This meant production of the civil Connie could pick up quickly and Lockheed maintained a two-year lead over their competitors. Douglas were now offering airlines their pressurised DC-6. In time they would develop it into the DC-7, arguably, the ultimate piston-engine airliner. Both aircraft were successful and popular with airlines; a total of 704 DC-6s were sold, followed by 336 DC-7s. Boeing, however, who were known primarily as a manufacturer of military aircraft, had gone in a different direction. They had decided to develop their B-29 bomber into a large,

The Boeing 377 Stratocruiser, developed from the B-29 bomber, was not popular with airlines and only 56 were built.

Douglas were now offering airlines their pressurised DC-6 (above), which they later developed into their DC-7 (below).

double-decked, long-range airliner, the 377 Stratocruiser. Plagued with reliability problems and burdened with high maintenance costs, the Stratocruiser was a failure for Boeing and only 56 were built. The company were saved from huge losses by large orders for the military tanker version but, unlike the DC-6, the Strat was no serious competitor for Lockheed's Connie.

While the former military Constellations would retain Lockheed's 049 designation, from now on, purpose-built civilian aircraft would be designated L-649 and L-749, with more powerful engines and more fuel in the outer wings. CAB certification was awarded on 14 October, 1945. Earlier that month, TWA had finally taken delivery of a civilian Constellation airliner followed by nine more in the following weeks. Their first transatlantic proving

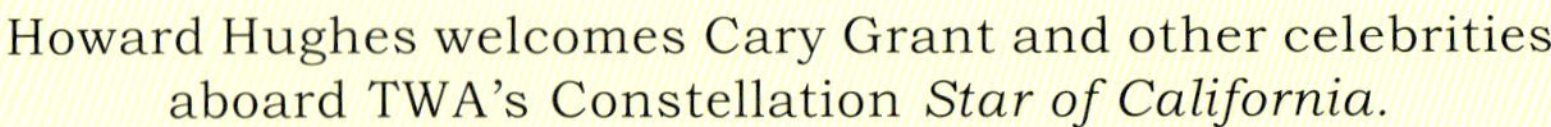
Howard Hughes welcomes Cary Grant and other celebrities aboard TWA's Constellation *Star of California.*

Howard Hughes at the controls of a Constellation.

flight left Washington, DC on 3 December 1945 and flew to Paris via Gander and Shannon. On 3 February 1946 TWA's Constellation *Star of California* began scheduled services from Burbank to New York, with Howard Hughes in command of the first flight. The non-stop flying time was 8 hours and 38 minutes, and, with a typical Hughes flourish, the passengers included Cary Grant, Walter Pidgeon, Veronica Lake, Linda Darnell, Edward G Robinson, Myrna Loy and Tyrone Power.

By late February 1946 they were operating regular flights from New York to Paris. For the first time passengers

SUPER CONSTELLATION
swissport

STAR OF SWITZERLAND
TOTAL
BREITLING

had access to affordable travel ($711 round trip) in fast pressurised aircraft which could cross the Atlantic above the weather. TWA had replaced Pan American as the airline of choice on their most prestigious route and TWA's traffic jumped 15% in a year. The glamorous profile of the beautiful Connie helped too; the DC-4 and Boeing Stratocruiser looked humdrum in comparison. It would be 16 months before Pan Am could level the score with Constellations of their own. Not to be outdone, they used the new plane to launch a round-the-world service, giving it the memorable flight number PA1. Their first flight, with Constellation *Clipper America,* departed in June 1947 and the PA1 would operate continuously until 1982.

A Pan American Constellation arrives at Shanghai during the first round-the-world service in 1947 (Pan Am Historical Foundation).

Pan American were the second customer for the Constellation, ordering ten 049s and thirty of the longer-range 349s.

John Wayne boards a TWA Constellation in 1955.

Queen Elizabeth II and Prince Philip leave a Qantas Constellation in 1954.

American Overseas Airlines became the third customer with an order for seven aircraft which they operated to Scandinavia, Iceland and Germany (both the airline and the aircraft would be taken over by Pan American in 1950). BOAC became the first European operator of the Constellation, with an order for seven. After the war, BOAC were having to make do with improvised and unsuitable aircraft such as the Avro Lancastrian, a civilianised version of the Lancaster bomber which carried a handful of passengers in a noisy, unpressurised bomb bay. It was then the policy of the British government that state-owned BOAC should buy only British-built aircraft, although they made an exception for the Canadian-built North Star (named 'Argonaut' by BOAC). Canadair had cleverly adapted the DC-4 by adding pressurisation and more powerful Merlin engines, producing a fast, over-the-weather aeroplane, but neither the Argonaut, nor any of the other aircraft in BOAC's immediate post-war fleet, were

An American Overseas Airlines Constellation at Oslo.

BOAC was the first European operator of the Constellation.

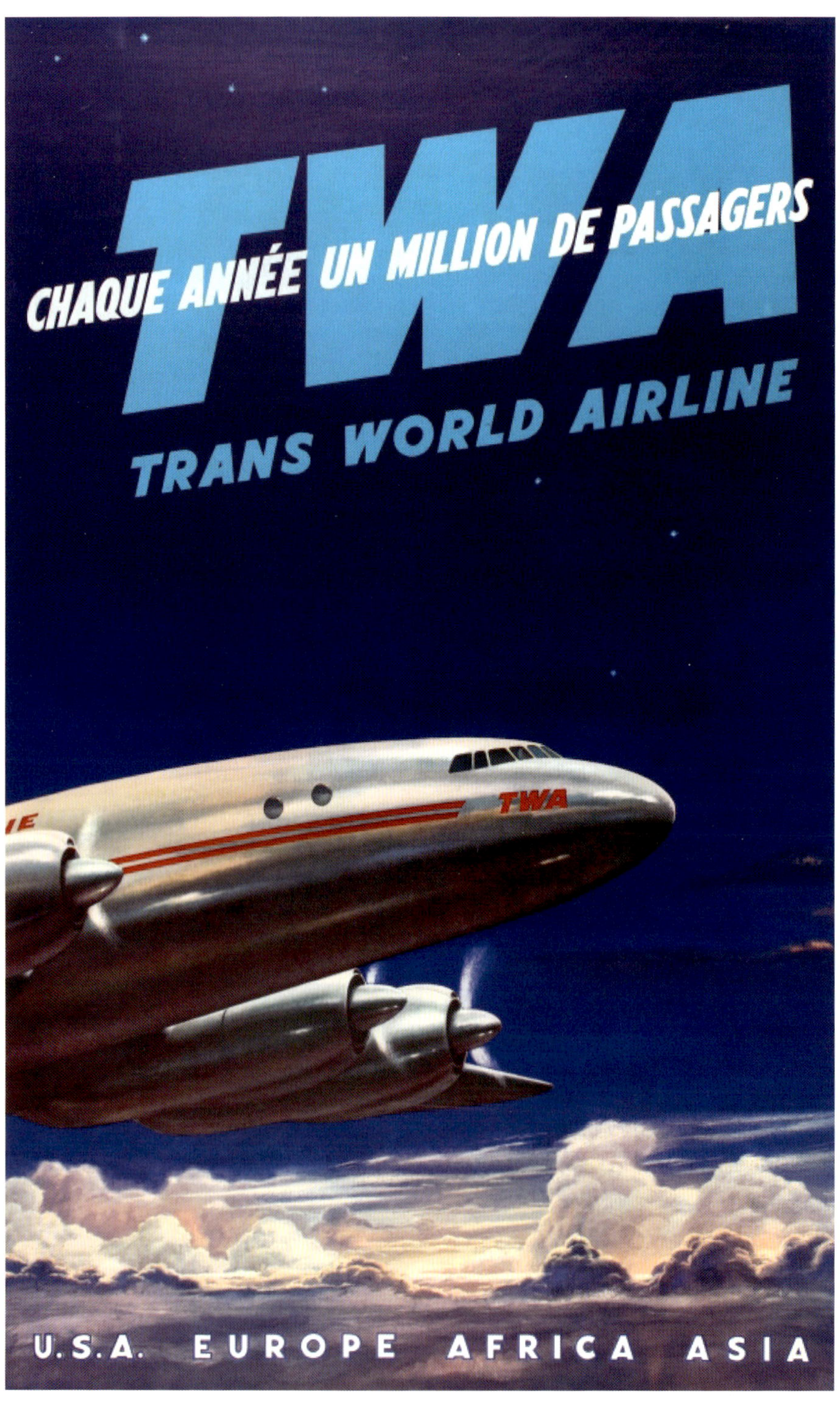
TWA
CHAQUE ANNÉE UN MILLION DE PASSAGERS
TRANS WORLD AIRLINE
TWA
U.S.A. EUROPE AFRICA ASIA

suitable for their crucial North Atlantic routes. Following a lengthy argument, which would resurface whenever BOAC tried to buy American planes, the British government allowed the corporation to order the Constellations and, a few years later, six Boeing 377 Stratocruisers.

CAPITO
N4903C

AIRWAYS

10 Early Setbacks

The Connie's entry into commercial service was not, entirely, a smooth ride. In June 1946, a Pan American Constellation suffered an uncontrollable fire in the number four engine shortly after take-off from New York. The fire was only extinguished when the engine entirely broke away from the aircraft, but the Connie was successfully belly-landed in a field with no injuries to passengers or crew. Pan Am had a reputation for excellent cabin service and, while the passengers huddled in blankets and waited to be rescued, the stewards served them hot tea and coffee along with a selection of sandwiches and cakes! The fire was later found to have been caused by the failure of a driveshaft for the cabin pressurisation pump, and modifications were made to the housing.

Tragically, there was a less happy result when a TWA Constellation, on a routine training flight from Reading, PA, experienced a major, inextinguishable cabin fire. Only one crew member survived the resulting crash, and he sustained lifechanging injuries. The subsequent enquiry threw up major design flaws in the electrical system. Electrical conductors known as 'through studs' (they allowed electrical current to pass through pressurisation seals) had been positioned so near to the aircraft's skin that arcing was occurring. Worse still, the studs were located close to high-pressure hydraulic lines carrying highly flammable liquid. Every Constellation in service was inspected and evidence of arcing was found in half of them. The CAB's report was damning. They slated the 'crudely deficient design' of the studs and found serious shortcomings

The wreckage of TWA's NC 86513, *Star of Lisbon.*

1944 Mahantongo Street. (Journal Staff Photo)

Constellation Planes Grounded for 30 Days Pending US Probe

Major Long-Range Aerial Carriers to Be Rechecked As Result of Crash At Reading in Which 5 Men Died

WASHINGTON, July 12 (U.P.)—International air travel was disrupted seriously today as the government grounded for at least 30 days all Constellation airplanes flown by United States Airlines.

The Civil Aeronautics Administration ordered the four-motored Constellations out of the skies pending an investigation of the causes of a crash of one of the planes near Reading, Pa., while on a training flight. Five persons were killed.

Pan American Airways, Trans-World Airlines and American Overseas Airlines, which used the planes, struggled with makeshift arrangements to handle hundreds of passengers who had reservations on the giant machines.

All three lines announced they would try to substitute Douglas DC-4 Skymasters for the Lockheed Constellations so that as few schedules as possible would be cancelled.

Approximately 42 of the aerial giants were in use by United States lines. The planes, largest land-based aircraft in commercial service, carry between 40 and 50 passengers each, and are one of America's major long-range carriers.

Hardest hit apparently was Pan American's Los Angeles-San Francisco to Honolulu service which carries 700 passengers a week. J. V. Roscoe, Pacific manager, said in San Francisco nine weekly round-trips will have to be cancelled until other planes can be submitted.

Civil Aeronautics Administrator T. P. Wright announced the 30-day "emergency suspension" of the

Cumbola Man Gets Appeal From Arrest

Alleging that his arrest and sentence to pay a fine of $100 and costs for allegedly resisting Anthony Lech, fish warden of Schuylkill Haven, who alleges that the defendant refused to show his fishing license on demand, John Shimkus, of Cumbola, was granted an appeal by Judge V. J. Dalton, who will hear the case in Court.

The defendant was engaged in fishing, it is alleged, when Lech demanded that he display his fishing license, and upon his alleged refusal to do so, he was hauled before Justice Allen Klahr, of Schuylkill Haven, and fined $100 and the costs, which amounted to $2.50, which he

Trieste Still Tense; Troops Patrol Border

BY JOSEPH BAICICH
United Press Staff Correspondent

Trieste, July 12 (U.P.)—American and British troops patrolled the entire Yugoslav occupation frontier across northeast Italy today in a bid to shut off smuggling of UNRRA foodstuffs to Slavs on general strike in riot-torn Trieste.

Allied Headquarters announced that many truckloads of food, mostly flour and at least partly of UNRRA origin, had been seized in the last few days at the Morgan line between the Yugoslav and Anglo-American zones of Venezia Giulia.

Authorities regarded the smuggling as a primary source for sustaining the spirit of violence in Trieste, where scarcely a day has passed lately without the cracking of some heads. At least 11 persons were wounded yesterday in rioting by 20,000 sympathizers of the Slovene strikers.

At least 20 persons had been arrested for smuggling UNRRA foodstuffs in to the strikers at Trieste. The frontier patrol was established in a move to choke off the line of supply to the strikers.

Trieste police charged that Slovenes were pouring into the city by the hundreds to join the fighting against the Italians. Alarmed Italian residents told officers that unless they were assured of protection, they would take the situation into their own hands—a threat of a step-up in the sporadic outbursts of the last few weeks.

Authorities announced last night that several persons were killed and in the fire detection and control systems. All Constellations were grounded for extensive modifications and it would be more than two months before they returned to the sky. Among the changes made were replacement of the Wright engines' carburettors with a fuel injection system. The Army had made the same modification to the engines of their B-29s, and it had significantly reduced engine fires. TWA's problems further worsened when, only a month after the Constellations had returned to the air, their 800 pilots went on strike. TWA's financial position had been weak before the strike and the relationship between Hughes and Frye began to deteriorate to the point where Frye's job was on the line. Another plane crash brought matters to a head.

The New York

TWA PLANES HALT IN U.S., OVERSEAS AS PILOTS STRIKE

Hughes had never lost his fascination with high-performance aircraft. His Hughes Aircraft company had built an experimental military reconnaissance plane, largely designed and test-flown by Hughes himself, called the XF-11. A few days before the Constellation was grounded, Hughes was piloting the prototype XF-11 on its maiden flight from Culver City Airport in California. He encountered serious control problems when one of the counter-rotating propellers failed and went unexpectedly into reverse. While attempting a forced landing on the Los Angeles Country Club golf course, he crashed into a house in Beverly Hills. His injuries were life-threatening, and he spent five weeks in hospital. Some of Hughes' biographers believe that the trauma of his wounds, along with the pain killers he became addicted to, played a role in the obsessive-compulsive personality disorder which later turned him into a recluse. The TWA pilots' strike occurred during Hughes' long recuperation and by 1947 Jack Frye had been fired.

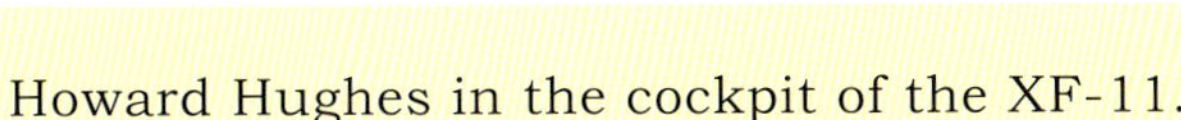

Howard Hughes in the cockpit of the XF-11.

Four times married, Frye would be killed at the early age of 54 (his car was hit by a drunk driver) while on his way home from work. Despite towering achievements, his death was largely ignored by the media — on that day the eyes of the world were on Buddy Holly, Ritchie Valens and the Big Bopper who had died in a Beechcraft Bonanza which crashed near Clear Lake, Iowa.[1] Howard Hughes, who was Frye's contemporary, would die seventeen years later aboard a Lear Jet taking him to hospital for kidney failure treatment. His reclusiveness and drug use had left him almost unrecognisable.

---

1 The young and inexperienced pilot of the hastily chartered Bonanza had become spatially disorientated on a dark night. Don McLean dubbed the tragedy 'The day the music died' in his 1971 song *American Pie*.

11 The Connie Spreads Her Wings

Despite these setbacks, the Constellation proved to be a popular aircraft with airlines both in the United States and overseas. Following the early orders from TWA, Pan American, BOAC and American Overseas Airlines, further Connies would be sold to KLM, Eastern Air Lines, Air France, Qantas, Lufthansa, Iberia Airlines, Panair do Brasil, TAP Portugal, Trans-Canada Air Lines, Aer Lingus, VARIG, Cubana de Aviación, Línea Aeropostal Venezolana, Pakistan International Airlines, Capital Airlines and Avianca. Total production, including military variants would come to 856

N749NL at Lelystad airport in Holland, 4 September, 2004, having been repainted in KLM's 1950's livery.

LEADERSHIP DEMANDS CONSTANT ACHIEVEMENT
TODAY'S MARK OF DISTINCTION
World-renowned as...
THE FINEST!
LOCKHEED
Constellation and Super Constellation
Airline passengers remember, with confidence, these facts of leadership of the Lockheed Constellation family of air transports:
FIRST with powerful Wright turbo-compound engines.
FIRST in scheduled non-stop transcontinental service.
FIRST pressurized transcontinental and transocean airliner in the 350 m.p.h. class.
FASTEST across the North Atlantic.
BIGGEST air transport in service—with room to accommodate 5 commodious separate cabins, including a luxurious club lounge.
UNSURPASSED comfort, luxury, decor and appointments—with interior specially created for the turbo-compound Super Constellation by world-renowned designer Henry Dreyfuss:
Most advanced air-conditioning.
Most air circulation per passenger minute.
Widest aisle in the roomiest main cabin.
More lavatory facilities.
UNSURPASSED in airline preference. A new Super Constellation airline starts service every month this year and into 1955.
UNSURPASSED in significant city-to-city timesaving to passengers—product of added speed.
Speed is useful only when it means practical timesaving, with at least equal dependability and comfort.
The turbo-compound Super Constellation offers unsurpassed practical, dependable, comfortable timesaving. When, with new design or power, significantly greater timesaving is practical, passengers will continue to . . .
FLY CONSTELLATIONS AND SUPER CONSTELLATIONS ON THESE 25 WORLD AIRLINES:
U.S.A.—Capital Airlines • Delta-C & S Airlines • Eastern Air Lines • Northwest Orient Airlines* • Pan American World Airways • Seaboard & Western* • TWA-Trans World Airlines.
NORTH & SOUTH AMERICA—Avianca (Colombia) • Cubana (Cuba) • LAV (Venezuela) • Panair do Brasil • Trans-Canada Air Lines • Varig* (Brazil).
EUROPE—Air France • B.O.A.C. (Great Britain) • Deutsche Lufthansa* (Germany) • Iberia* (Spain) • KLM (Holland) • Portugal*.
ASIA & AFRICA—Air India • El Al Israel • Pakistan International • Thai Airways* (Thailand) • South African Airways.
AUSTRALIA—Qantas Empire Airways. *Soon
LOCKHEED
AIRCRAFT CORPORATION • BURBANK, CALIFORNIA, AND MARIETTA, GEORGIA
LOOK TO LOCKHEED FOR LEADERSHIP
StarRunn
1954

B·O·A·C
G-ALAL

TRANS-PERUANA
TP
OB-R-915
Skyways of London

FLY TWA TO
NEW YORK
TWA
TWA
820
N6020C
TWA
TRANS WORLD AIRLINES

aircraft. TWA, alone, would at various times operate a fleet of 188 Constellations of varying marques, including some second-hand aircraft (not all were in service at the same time).[1] By 1947 five different airlines were operating the Constellation across the North Atlantic. The success and popularity of the Constellation forced Douglas to quickly

1 *Lockheed Constellation* by Curtis K Stringfellow and Peter M Bowers. Motorbooks International, 1992.

develop their unpressurised DC-4 into the bigger, faster, pressurised DC-6 and, ultimately, the DC-7.

As the first widely used pressurised airliner, passengers loved being above the weather in a Constellation. The turbulence which was inevitable in earlier unpressurised aircraft had often caused airsickness and was a major deterrent to flying. Because pressurisation was so new, it took Lockheed a while to overcome cabin temperature problems. Stewardesses became known as 'Hot or Cold Running Hostesses', so often did they come to the flight deck with pleas from passengers for cabin temperature changes.

Lockheed developed the Constellation continually throughout its history, and the maximum gross weight of the 049 soon grew from 86,250 pounds for the original 049 to 98,000 pounds for the 049E. Customers were also offered the options of Pratt & Whitney R-2800 engines, or British Bristol Centaurus engines, but there were no

Lockheed offered the Bristol Centaurus as an alternative powerplant, but there were no takers.

This TWA Constellation overran the runway during a bad weather landing at Chicago Midway. There were no injuries to passengers, crew or bystanders.

A-C-E

takers for the alternative — every customer stayed with the original Wright powerplants.[2] The windshield sticker prices, however, ranged from $685,000 to $720,000, a far cry from the $405,000 that Lockheed had originally estimated.

By May 1946, following a request from Eastern Air Lines, Lockheed was already working on a much-improved variant. The Model 649 would have more powerful engines, improved cabin heating, air conditioning and ventilation, along with a significant increase in maximum take-off weight. Other modifications included better engine cowlings, reversible pitch propellers and increased flap deflection. Cabin noise and vibration were reduced by multiple layers of fibre glass insulation and shock-mounted wall panels while fire-resistant fabric was introduced. The interior skin would now be connected to the outer skin by rubber mounts to further reduce vibration and small wide-angle lenses were installed in the cockpit so that the engineer could see the engines in flight. The new Connie was considerably faster and more economical — launch customer Eastern Air Lines named it their

2 *Lockheed Constellation* by Curtis K Stringfellow and Peter M Bowers. Motorbooks International, 1992.

Eastern's New-Type Constellation. World's largest, fastest, most powerful Airliner

Improvements from the Model 649 onwards included reversible pitch propellers, better engine cowlings and increased flap deflection. Below: a Lockheed L-749 Constellation of TWA with Speedpack.

'Gold Plate Constellation'. The 649 Constellation was, in time, upgraded to the 749 with further improvements including outer wing tanks that carried an additional 1,555 gallons and an improved exhaust system. Most of the airlines who had ordered 649s upgraded their orders to the 749 model.

The freight-carrying capacity of the Connie was hugely increased by the option of a Speedpack: a cleverly designed under-fuselage freight container which was developed at the request of Eastern, and which could carry up to 395 cubic feet, or up to 8,300 pounds, of cargo. Testing showed that the decrease in airspeed was only 12 mph and, Lockheed claimed, the Speedpack could be attached or detached from the aircraft in only two minutes. The TWA strike, however,

KLM Constellation with Speedpack attached.

The Speedpack cargo pod from a Qantas Constellation (Qantas Heritage Collection).

led to cancellation of their orders for the 649, leaving Eastern as the only customer to receive it.

By 1952 BOAC had begun operating de Havilland's Comet 1 jet on their African and Asian routes, but the early Comets lacked the range for transatlantic flights. The Comet 1 was withdrawn from operation in 1954 following a string of accidents, but jet airliners would return to airline service in 1958 with the Comet IV and Boeing 707. So successful and popular was the Constellation, however, that she remained in scheduled passenger service for a full decade after the introduction of jets, while military and freighter versions flew for much longer. One reason for the Constellation's longevity was her remarkable fuel efficiency. A 2005 report by the Dutch National Aerospace Agency showed that modern jet aircraft, such as the A380, had only just matched the fuel efficiency of the best piston-engine aircraft, especially the Constellation, which was at least twice as fuel-efficient as the first jet-powered aircraft.[1]

1 *Fuel efficiency of commercial aircraft: An overview of historical and future trends.* Dutch National Aerospace Agency, 2005.

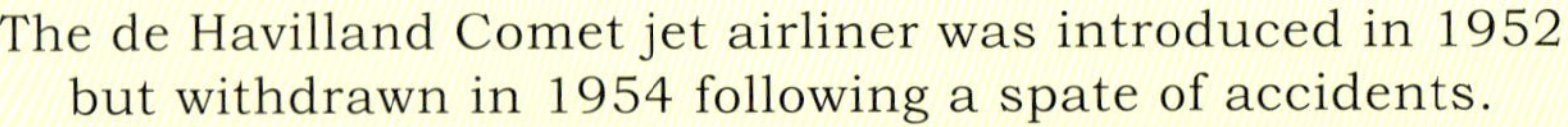

The de Havilland Comet jet airliner was introduced in 1952 but withdrawn in 1954 following a spate of accidents.

By 1958 the Comet IV and Boeing 707 had entered airline service, but so succesful was the Constellation that she remained in scheduled passenger service for a full decade after the introduction of jets, while military and freight versions flew for much longer.

Modern aircraft like the A380 have only recently matched the fuel efficency of the Constellation, which was twice as fuel-efficient as the first jets.

The popularity of the Connie, however, was based on more than her speed, range and service ceiling. Passengers loved her (and still love her) for her style. When Frank Sinatra released his album *Come Fly With Me* in 1958, it was a Constellation that he cocked his thumb towards. The period from the end of the Second World War until the advent of wide-bodied aircraft in the late 1970s, became known as flying's Golden Age. Americans were enjoying a post-war economic boom. Despite the earlier fears of economists, American industry quickly adapted to peace, returning veterans found well-paid work and disposable incomes rapidly increased. It has been calculated that by 1945 Americans were saving 21% of their personal disposable income, compared with just 3% in the 1920s.[2] Gross national product soared from $200 billion in 1940 to $300 billion by 1950. People bought new homes and filled them with new appliances.

2 *A Consumer's Republic. The Politics of Mass Consumption in Post-war America,* by Lizabeth Cohen, Vintage Books, 2004.

Frank Sinatra's 1958 album *Come Fly With Me* featured TWA Constellations on the cover.

At the same time, they began to travel across the vast continent of North America in ways that few could have dreamed of before the war. The Lockheed Constellation, which was a quantum leap over its predecessors, came into service at exactly the moment that Americans were demanding faster and more reliable air connections to take them between their major cities.

It was an era in which people dressed to fly, friends and relatives came to the airport to see them off and airlines lavished multi-course meals and highly attentive service on their customers. The advertising industry quickly saw

G-ANTF

ACE FREIGHTERS

that the glamorous Connie could be a marketing icon and rarely missed any opportunity to show off her film star good looks. Movies of the period, including *How to Marry A Millionaire* and *It Started With a Kiss*, often showed leading characters arriving or departing by Constellation, and celebrities were constantly photographed boarding or leaving her. Best of all, a Connie became the personal choice of the President.

The glamorous Connie quickly became a movie star. Opposite: a French movie based on the loss of an Air India Constellation. Above: Jane Russell is photographed aboard a Constellation.

Columbine II, a specially adapted VC-121E, was President Eisenhower's personal transport, and the first aircraft to carry the call sign Air Force One.

Columbine III, a modified Super Constellation was the last piston-engined plane to be used as primary air transport by an American president.

In 1953 a military Constellation, designated VC-121E, was specially adapted to become President Dwight D Eisenhower's personal transport. It was named *Columbine II* after the official flower of Colorado, the adopted home state of Mamie Eisenhower (*Columbine I* had been General Eisenhower's personal transport when he was Supreme Commander of Allied Forces in Europe during World War Two). Legend has it that the call sign 'Air Force One' was first used by the pilot of the *Columbine II* during a flight to Florida. He was worried that air traffic controllers might confuse the presidential plane's call sign, Air Force 610, with a similar call sign of a nearby commercial airliner. *Columbine II* remained in service until 1954 when it was replaced by *Columbine III*, a modified Super Constellation that was longer, faster, and more comfortable than its predecessor. *Columbine III* was the last piston-engined plane to be used as primary air transport by an American president. She was eventually retired from the Air Force in 1966 and is today preserved at the National Museum of the United States Air Force at Wright-Patterson Air Force Base near Dayton, Ohio. *Columbine II* and *III* both became media stars and their prominence helped enhance the Connie's already fabled status.[1]

1 *Columbine II* was purchased by Dynamic Aviation in 2016 and taken to Bridgewater, Virginia for restoration.

12 New Sales to the Military

During a low point in sales for the civilian versions, Lockheed began to receive significant government orders for the Model 749 Constellation (now designated C-121 for the military version): ten for the USAF and two for the Navy (the naval versions were named 'Warning Stars'). The military view of the aeroplane had changed considerably. Although they had originally seen it as cutting-edge technology which could lead to vast improvements in the speed and range of their airlifters, they now saw the Connie as a tried and tested workhorse that presented little risk. Additionally, Douglas had finally caught up with Lockheed and their DC-6 now offered similar levels of performance. The Air Force responded by ordering 101 DC-6As.

A C-121 of the Blue Angels, the US Navy's aerobatic display team.

Above: US Navy C-121 Warning Stars.
Below: General MacArthur's C-121 *Bataan* (Tom Beudeker).

The C-121 could carry either heavy cargo, medical evacuees, or passengers. The floor was reinforced, and a cargo door was added while the gross weight was increased to 107,000 lbs. These aircraft would play a key role during the Berlin Airlift (1948–9); they didn't fly directly to Berlin, but covered nearly six million miles transporting goods across the North Atlantic during the first month. General Douglas MacArthur used a C-121A named *Bataan* during the Korean War. This aircraft was eventually taken over by NASA and used at the Goddard Space Flight Center in Greenbelt, Maryland. Many of these military Connies would eventually find civilian work as crop sprayers or cargo planes.

The Navy had a different use for the Connie. They equipped them with radar and used them as early warning scouts. Airborne Early Warning (AEW) Radar was then an important new concept as Cold War tensions increased.

K

13 The Super Connie

Test flights had shown that the Constellation could be flown with an all-up-weight up to 30,000 pounds greater than that of the 749, proving to Lockheed that the original design had the potential to be significantly lengthened. Detailed design work on a stretched version — the 1049 — quickly dubbed the Super Constellation, began in 1949. Instead of building a new prototype, Lockheed bought back the first C-69 aircraft from its current owner, who happened to be Howard Hughes!

A Super Constellation being refuelled in 2016.

A range of Lockheed Constellation types, including the L-049, L-749, L-1049 and L-1649, became Air France's flagship brand on its main long-haul routes. No fewer than 62 were used in the fleet between 1946 and 1967.

Kelly Johnson with a Super Constellation.

The fact that Douglas had closed the gap on Lockheed and were now selling their DC-6 in large numbers, would have been a considerable spur for Lockheed to once again 'up the ante' with a bigger, better aeroplane. The Connie's fuselage was stretched by 18 feet 4¾ inches, bringing the total length to 113 feet 7 inches. For comparison, the Boeing 707 is 152 feet long, only a third longer, which underlines what an ambitious stretch this was. This was accomplished by putting two additional, equal-length 'plugs' into the fuselage, fore and aft of the wing. New Wright engines were used, producing 2,700 hp each, and the fuel capacity was increased to 6,550 gallons. This was achieved by adding a 730-gallon tank to the centre section, which increased the range by 500 miles. Larger square windows replaced the original round ones, increasing visibility for passengers by 85%. In the cockpit, the old nine-window arrangement was replaced by seven windows with a 21% increase in the pilots' view, along with an additional 7 inches of headroom. The longer fuselage required bigger rudders, and they were enlarged by the addition of a straight section within the previously

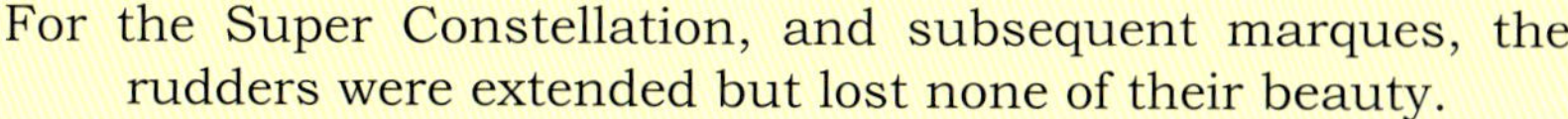

For the Super Constellation, and subsequent marques, the rudders were extended but lost none of their beauty.

fully elliptical tail fins. This was so skilfully incorporated above and below the horizontal stabiliser, that the tail lost none of its unique 'Connie' beauty.

Lockheed L-749-79 Constellation *Jose Marti* of Línea Aeropostal Venezolana. This aircraft was lost in 1956 with no survivors.

Eastern Air Lines L-1049C Super Constellation.

FLEVOLA
LE

K·L·M
LE
PH-XIV
DROME
VERBODEN TE BETREDEN A.U.B!
DAF
Daf 600 prototype 1958

The redesign was an opportunity to fix many of the problems that had beset the earlier marque, and a total of 550 improvements were made. The final aircraft was certified at a maximum gross weight of 120,000 lbs, 4,000 lbs more than had originally been guaranteed. Eastern Air Lines became the first customer when they signed an order for ten in April 1950 and the prototype made its first flight successfully on October 13, 1950. By December TWA had ordered a further ten.

The 1049A and 1049B models fell somewhat short of Lockheed's intended performance goals, which weren't fully achieved until they produced the 1049C. This was accomplished by using turbo-compound engines which had three power recovery turbines on each engine. The exhaust from six of the cylinders was fed back to the engine via a fluid coupling, increasing the engine's power by 20%. Lockheed were now, once more, ahead of Douglas — beating their new DC-7 (which used the same turbo

An L-1049G Super Constellation of TWA.

QANTAS
presents the Super Constellation
QANTAS
AUSTRALIA'S OVERSEAS AIRLINE

compound engine) by three months. Sadly, the turbo compound engines proved to be unreliable and inflight shutdowns became too frequent. So much so, in fact, that pilots nicknamed the Super Connie 'the world's best three-engined aircraft'. The main problem was excessive exhaust flaming which burned away the hoods of the power recovery turbines. Exhaust flames often reached beyond the trailing edge of the wing. Not only did this threaten the aircraft's structure, but it also unnerved the passengers, especially at night! It took $2 million in research and months of testing to find the solution, which was a two-inch wide ring of armour around each turbine.

The more powerful engines increased the cruising speed to 330 mph but, at the same time, decreased the range by a significant 300 miles. On the plus side, however, improved insulation and soundproofing made the Constellation one of the quietest airliners of her day.

Excessive exhaust flaming was a complex problem which cost $2 million to solve.

An L-1049F with cowling open.

The 1049C was sold to seven airlines: Eastern, Air France, KLM, Trans-Canada, Qantas, Pakistan International and Air India. Only four 1049Ds were built, but the 1049E was sold to Qantas, KLM, Avianca, Iberia, Trans-Canada, Air India and Cubana.

The 1049F was a freight version, but the 1049G was a response to what was now seen as a deficiency in range. The 1049G was offered with optional wingtip tanks which increased the range by a further 700 miles, along with another 107 improvements. The wingtip tanks did more than extend the range; they also changed the 'look' of the Connie, whose wings now appeared to be less elliptical and more streamlined than before. Opinions differ as to whether it was an aesthetic improvement or not! The 104G was the most popular of all the variants and 102 were sold: forty-two for US airlines and fifty-nine for foreign airlines. The range of the 1049G was 4,140 miles with a

The L-1049F (below and opposite) was, originally, a freight version (opposite top Christian Waser).

limited payload, but this was still not enough to fly from New York to Paris reliably and with a reasonable payload.

The last of the Super Constellation series was the 1049H. This was a 'convertible' aircraft which could be quickly changed from an all-passenger configuration to all-cargo. This variant also proved to be popular, and a total of 53 were built for delivery to 15 airlines and one leasing organisation. When production of the Super Connie finally ended, 259 had been built for a wide range of civil operators. Flying Tigers bought the last one from the production line in November 1958, and the last to have been unsold at the time production ended was finally delivered to Slick Airways in September 1959.

Wingtip fuel tanks changed the look of the Connie.

The Super Constellation was popular with the United States Navy too. The prototype, which had been nicknamed 'Old 1961', was fitted with a wide range of radomes for exhaustive flight testing. This resulted in the PO-2W Warning Star variant, of which the Navy ordered 244 along with a number of cargo-transport variants designated R7V-1s. The United States Air Force also operated numerous variants of the Super Constellation. For details of all the various military types, see the data on page 175.

PARCEL FORCE
4
MILITARY AIR TRANSPORT

RVICE
MATS
Atlantic

14 The Ultimate Constellation

By the mid-1950s Douglas had considerably improved the range of their DC-7 with the 'C' variant, which became known as the Seven Seas. This was the first airliner to have genuine transatlantic range and it prompted Lockheed to look again at the Constellation and consider yet another upgrade. In May 1955 they began work on the 1649A by putting an entirely new wing onto the 1049G, along with major modifications to the fuselage, tail and powerplants. The new wing had a span of 150 feet — the old elliptical design was now replaced by a much higher aspect ratio (12:1) with moderate sweepback and thinner airfoils.

The Douglas DC-7C Seven Seas was the first airliner with genuine transatlantic range.

The passenger cabin became quieter with the engines mounted five feet further outboard, and propeller noise was further reduced by lower gearing and new propellers. Another 900 lbs of soundproofing gave passengers a much more comfortable ride and the fuel capacity was hugely increased from 4,760 gallons to 9,278 gallons. The Connie's range was now well ahead of the DC-7 with more than transatlantic capability; even transpolar flights, including Los Angeles to London, were now possible. The new plane would be called the Starliner, in keeping with Lockheed's tradition of naming their aircraft after celestial bodies.

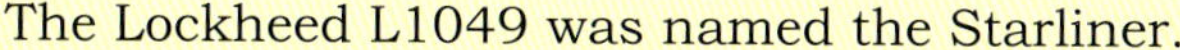

The Lockheed L1049 was named the Starliner.

The first 1649A, with the registration number N1649, flew on October 10, 1956. Test flights showed an increase of speed up to 377 mph, and a cruising speed of 290 mph. The range was 4,940 miles, which could be extended up to 6,180 miles with a reduced payload and the price per aircraft was now $2.9 million.

The new aircraft had the greatest range and highest speed of any contemporary airliner, including the DC-7. The Starliner could fly from New York to Paris in three hours less than the DC-7C and every European capital was now within nonstop range from New York. TWA order 25, later increased to 34. Air France ordered ten and promptly broke the speed record from Los Angeles to Paris (seventeen hours and eleven minutes). Further orders followed from Lufthansa, Alitalia and VARIG, while the United States Navy ordered an AEW version, although this was later cancelled due to budget restraints. The range and speed of the Starliner allowed airlines to pioneer new nonstop routes that could not then have been contemplated with any other aircraft, but not for very long. By 1958, the de Havilland Comet IV and Boeing 707 jet airliners would enter service and journey times would halve.

The Lockheed Starliner was, beyond any doubt and by any measure, a superb aircraft — arguably the best of all the piston airliners including the DC-7, but this ultimate version of the Constellation came into service too late to find a large market. The last aircraft built was delivered in 1958, sixteen years after the first Connie and marking the longest production run of any piston-engine airliner. By 1960, however, the Constellation was fading into history, although she would survive as a freighter and charter aircraft for much longer. A crucial chapter of aviation history had closed, but, at the same time, an icon had been created that would live long beyond those who had made the beautiful Connie possible.

BREIT
NG SUP

HB-RSC
HB

15 Constellations in History

In 1968, in London, Irish businessman Aodogán Ronan O'Rahilly announced plans for a 'pirate' television station to be known as Caroline TV. The new broadcaster's programmes would, he claimed, be transmitted from two Lockheed Constellation aircraft which would take turns to orbit above the North Sea. O'Rahilly had already turned British broadcasting upside down by running the best known of several 'pirate' radio stations which had been broadcasting pop music to British listeners, illegally, from ships outside British territorial waters. His 'pirate' radio stations had found a big audience during the preceding four years, and even the staid BBC was eventually forced to launch a new station, Radio One, to broadcast popular music nonstop.

The owner of 'pirate' radio station Radio Caroline planned to broadcast 'pirate' TV from Lockheed Constellations.

By 1968 new legislation had forced the closure of all the pirate radio stations but the new laws had not entirely excluded the possibility of pirate television. O'Rahilly planned to fly his Constellations at 20,000 feet, twelve miles off the east coast of Britain from where he would beam television every night from 6.00 pm to 12 midnight five days a week and between 6.00pm to 2.00 am on the other two days. Programmes included a two-hour daily pop show, a chat show called 'Out of Your Mind', feature films, documentaries, series, cartoons and news. Each plane was to have an on-board studio for news broadcasts and live interview inserts. Although the concept seemed farfetched, American Forces had successfully used a similar system, known as Stratovision, to broadcast television signals to Vietnam. Over £1.25 million was said to have been invested in Caroline TV and sufficient advertising promised to keep the station on the air for at least six months, but no broadcasts were ever made, and the project soon faded away.

Stratovision had been used succesfully to broadcast TV from B-29s to Vietnam.

In the early 1970s, the Rhodesian bush-pilot, gunrunner and sanctions buster, Jack Malloch, operated an ex-VARIG L-1049 Super Constellation which flew arms and other supplies into Rhodesia in defiance of United Nations sanctions. When the aircraft was grounded for lack of spares, it became a clubhouse at Charles Prince Airport near Salisbury, Rhodesia.

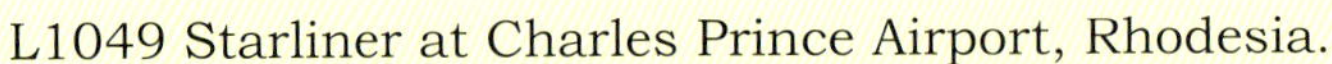

L1049 Starliner at Charles Prince Airport, Rhodesia.

16 The Airlines

TWA was both the launch customer and the biggest single commercial operator of the Lockheed Constellation. Some might also argue that the classic 1950s, red TWA livery was the finest looking of them all. They received their first aircraft on October 1 1945, and launched their first Constellation service, from Washington, DC to Paris via Gander and Shannon, just five weeks later. With the Second World War over, the aeroplane finally delivered what Hughes and Frye had always hoped for: a clear edge over their competitors. TWA's image became as distinct in the minds of passengers as those of Pan American, United and American; it was a huge step forward for an airline with such small beginnings, and the Constellation played a major role.

TWA's inaugural Constellation flights on key routes including New York to Paris and Rome, and Los Angeles to New York, were often loaded with movie stars and they gained TWA huge publicity. Some of the kudos was lost when early accidents caused the aircraft to be grounded for modifications, but not all of it. The public took both the Connie and TWA to their hearts, and the airline would remain one of America's most important and influential until it ran into financial problems in the early 1990s.

The second commercial order for Constellations came from Pan American, who ordered 40 as soon as Lockheed's agreement with Howard Hughes allowed them to. Juan Trippe was known for sponsoring new aircraft himself, rather than following in the wake of another airline, and it must have galled him to see Hughes and TWA not only receive their Constellations before Pan Am, but also to be awarded crucial international routes by the Civil Aeronautics Board. When TWA successfully launched the Constellation into international service, Pan Am lost their unofficial title of the 'Chosen Instrument', on some of their most prestigious routes.

The first of the Pan Am Connies, with 54 seats, was delivered on 5 January, 1946 and christened *Clipper Mayflower.* A second arrived a week later and they had begun North Atlantic Constellation services by 14 January 1946. This showed what a strenuous effort they made to catch up, because TWA itself did not start scheduled transatlantic services until 5 February. Pan American had taken delivery of 22 Model 049 Constellations by the end of May 1946. Two went directly to Panair do Brasil, still a Pan Am subsidiary, and they would receive eleven more during the 1950s as they were retired by Pan Am.

On 17 June 1947 one of Pan Am's four Constellation Model 749s, made the first round-the-world airline flight from New York to San Francisco. *Clipper America* left from La Guardia Field in New York. Making en route stops in Gander, Shannon, London, Istanbul, Dhahran, Karachi, Calcutta, Bangkok, Manila, Shanghai, Tokyo, Guam, Wake, Midway and Honolulu, the Clipper arrived in San Francisco on 29 June. As Pan American did not have authority to operate domestic flights within the United States, the Clipper flew empty to New York, arriving at La Guardia on 30 June, via Chicago, to complete the journey. Later, when Pan Am took over American Overseas Airlines (AOA), seven more 049s were added, bringing the total Connie fleet to 33. But the Connie did not remain

in service with Pan Am for very long. By the end of the 1950s, the entire Constellation fleet had been sold either to Panair do Brasil, Cubana, Delta, or Air France.

The primary operator of the L-649 Constellation was Eastern Air Lines who ordered 22, with the first being delivered in May 1947. Eastern had modified their original order for 14 L-049 Constellations to 14 L-649 Constellations. Eastern was the only airline to receive the L-649 straight from the production line, as TWA and Air France converted their L-649 orders into the improved derivative, the L-749, which was to be produced alongside the L-649, although, as it turned out, the L-749 finally entered service a month before the L-649.

The British Overseas Airways Corporation (BOAC) had faced a chronic problem in the immediate aftermath of the Second World War. There were no British-built aircraft that could satisfactorily serve their crucial North Atlantic routes. After a protracted argument, the British government finally allowed them to buy six Constellations. They were later able to add five new Lockheed 749 Constellations from Aerlínte Éireann in Ireland, without dollar expenditure, which enabled BOAC to serve Australia with Constellations from 1949. When the Comet 1 was grounded in 1954, following a series of accidents, BOAC were forced to buy back Constellations on the second-hand market at inflated prices. A total of 25 Constellations were operated by BOAC at various times, including 12 749As obtained from Capital Airlines in the mid-1950s, with BOAC's older 049s in part exchange.

TWA
TWA
TWA
TWA

TWA
TWA

17 Twilight of the Constellation

With the arrival of jet airliners in the late 1950s and early 1960s (including the de Havilland Comet IV, Boeing 707, Douglas DC-8, Convair 880, and Sud Aviation Caravelle), the Constellation became obsolete. TWA's last flight of a Constellation was made by an L-749 on May 11, 1967, from Philadelphia to Kansas City, Missouri; while the last scheduled passenger flight in North America was by Western Airlines' N86525 in Alaska, Anchorage to Yakutat to Juneau on 26 November 1968.

Constellations continued to carry freight for many years and were being used by Eastern Air Lines' shuttle service between New York, Washington, D.C., and Boston

N86525 made the last scheduled passenger flight of a Constellation in North America in 1968. Here she is seen at Chicago Midway in 1952.

as late as 1968. An Eastern Air Lines Connie still holds the record for a New York to Washington, DC flight from take-off to touchdown in just over 30 minutes. The record was set prior to speed restrictions by the Federal Aviation Administration (FAA) below 10,000 feet.

Lockheed did not build a large passenger aircraft again until the L-1011 TriStar which first flew in 1972. Although it was widely recognised to be a technically superb aeroplane, the TriStar was a commercial failure, and Lockheed left the commercial airliner business permanently in 1983.

The L-1011 TriStar was Lockheed's last commercial airliner (Jan Ostrowski).

18 The Survivors

While more than 50 Constellations survive in varying states of restoration, only one, sadly, is currently in flying condition. VH-EAG, a C-121C, which was restored in Tucson, Arizona by the Historical Aircraft Restoration Society, was flown to Australia in February 1996. At the time of writing, the aircraft remains airworthy and is an active participant in the Australian airshow circuit.

VH-EAG 'Southern Preservation' Lockheed C-121C
Super Constellation (L-1049F).

AIR FRANCE
AMÉRIQUE DU SUD

237
N6237G

237
SHELL

19 Tally of Production Aircraft

Below, are the full details, including the countries, registration numbers (second and subsequent registrations in brackets), models, construction numbers and operators, for Lockheed's entire production of civilian and military Constellations of all variants.[1]

Registration	Model	C/N	Remarks
Pakistan (AP)			
AP-AFQ/RFS	1049C-55	4520/4522	Pakistan Gov' (3)
AP-AJY, RSZ	1049H	4835,4836	PIA; to Indonesia AF
China (Taiwan, B)			
B-1809 (2)	1049H	4853	
Chile (CC)			
OC-CCA (5)	049-46-59	2069	
Canada (CF, C)			
CF-AEN (2)	1049H	4821	
CF-8D8 (2)	1049H	4809	To C-FBDB 1975*
CF-BFN (2)	1049H	4825	
F-NAJ, NAK (2)	1049H	4828,4829	
F-N AL, NAM (2)	1049H	4831,4832	
CF-PXX (3)	1049G-82	4580	From 1049E-55-01
CF-RNR (3)	1049C-55	4544	
CF-TEV 1049G-82	4641		Trans-Canada
CF-TEV 1049G-82	4643		Trans-Canada
CF-TEW (2) 1049G-82	4682		
CF-TEX (2)	1049G-82	4683	
CF-TEX/TEZ	1049H	4850,4851	Trans-Canada
CF-TGA/TGE	1049C-55	4540/4544	Trans-Canada (5)
CF-TGF/TGH	1049E-55	4563/4565	Trans-Canada (3)
CF-WWH (2)	1049H	4820	
C-GXKO (2)	C-121A	2601	Ex48-609
C-GXKR (2)	C-121A	2604	Ex 48-612
C-GXKS (2)	C-121A	2609	Ex 48-617
Morocco (CN)			
CN-CCM (2)	749-79-22	2515	
CN-CCN, CCO (2)	749-79-46	2675, 2676	
CN-CCP (2)	749A-79-46	2627	
CN-CCR (2)	749-79-22	2512	

1 Data from: *Lockheed Constellation* by Curtis K Stringfellow and Peter M Bowers. Motorbooks International 1992.

Registration	Model	C/N	Remarks
Bolivia (CP)			
CP-797 No. 1 (4)	749A-79-32	2548	
CP-797 No. 2 (3)	1049H-82	4807	
CP-797 No. 3 (5)	1049H-82	4801	
PORTUGAL (CS)			
CS-TLNTLC	1049G-82	4616/4618	TAP (3)
CS-TLD (2)	1049H-82	4808	
CS-TLE (5)	1049G-82	4677	
CS-TLF (3,5)	1049G-82	4672	
Cuba (CU)			
CU-C-601, 602	1049G-82	4632, 4633	Cubana CU-C-631
1049G-82	4675		Cubana
CU-P-573	1049E-55	4557	Pasquel, leased to Cubana
CU-T-547 (2)	049-46-26	2036	
CU-T-601 (2)	1049G-82	4632	Changed from CU-C-601
CU-T-602	1049G-82	4633	Changed from CU-C-602
CU-T-631	1049G-82	4675	Changed from CU-C-631
Uruguay (CX)			
CX-BBM (3)	749A-79-33	2641	
CX-BBN (5)	749A-79-33	2661	
CX-BCS (3)	749A-79-33	2640	
CX-BEM (2)	1049H	4818	
CX-BEN			Model unknown
CX-BHC (4)	749-79-31	2565	
CX-BHD (3)	749-79-82	2548	
CX-BGP (2)	749A-78-52	2668	
Germany (D)			
D-ALAK	1049G-82	4602	Lufthansa
D-ALAN	1649A-98	1040	Lufthansa
D-ALAP	1049G-82	4637	Lufthansa
D-ALEC	1049G-82	4640	Lufthansa
D-ALEM	1049G-82	4603	Lufthansa
D-ALER	1649A-98	1041	Lufthansa
D-ALID	1049G-82	4647	Lufthansa
D-ALIN	1049G-82	4604	Lufthansa
D-ALOF	1049G-82	4642	Lufthansa
D-ALOL	1649A-98	1042	Lufthansa
D-ALOP	1049G-82	4605	Lufthansa
D-ALUB	1649A-98	1034	Lufthansa
SPAIN (EC)			
EC-AIN/AIP	1049E-55	4550/4552	Iberia (3)
EC-AMP	1049G-8L	4673	Iberia
EC-AMQ	1049G-8L	4676	Iberia
EC-AQL (5)	1049G	4553	From 1049E-55
EC-AQM/AQN (3)	1049G-82	4644, 4645	
EC-ARN	1049G-8L	4678	
EC-BEN	1049G-82	4519	From 1049C-55-81 VIA E-55-01

Registration	Model	C/N	Remarks
Eire (Ireland, El)			
EI-ACR, ACS	749-79-32	2548, 2549	Aerlínte Éireann
El-ADA	749-79-32	2554	Aerlínte Éireann
El-ADD	749-79-32	2555	Aerlínte Éireann
El-ADE	749-79-32	2566	Aerlínte Éireann
EI-ARS (4)	1049G-82	4672	
Ethiopia (ET)			
ET-T-35	C-121A 2608	48-616	to Ethiopian Airlines
France (F)			
F-BAZN/BAZD	049-46	2072/2075	Air France (4)
F-BAZE/BAZH	749A-79-46	2624/2627	Air France (4)
F-BAZI/BAZK	749-79-22	2513/2515	Air France (3)
F-BAZL	749-79-22	2538	Air France, to 749A-79-46
F-BAZM/BAZO	749-79-22	2545/47	Air France (3) to 749A-79-46
F-BAZP	749-79-22	2550	Air France to 749A-79-46
F-BAZQ	749-79-22	2512	Air France to 749A-79-46
F-BAZR (4)	749-79-22	2503	
F-BAZS, BAZT	749A-79-46	2628, 2629	Air France
F-BAZU, BAZV (2)	749A-79-46	2525, 2526	Air France, from 749-79-22
F-BAZX, BAZY	749A-79-46	2527, 2528	Air France, from 749-79-22
F-BAZZ	749A-79-46	2674	Air France
F-BBDT	749A-79-46	2675	Air France
F-BBDU, BBDV	749A-79-46	2676, 2677	Air France
F-BGNN/BGNJ	1049C-55-81	4510/4519	Air France (1O)
F-BHBN/BHBH	1049G-82-98	4620/4627	Air France (8)
F-BHBI	1049G-82	4634	Air France
F-BGBJ	1049G-82	4639	Air France
F-BHBK	1649A	1011	Air France*
F-BHBL	1649A-98-11	1020	Air France
F-BHBM, BHBN	1649A-98-11	1027, 1028	Air France
F-BHBO/BHBQ	1649A-98-11	1031/1033	Air France (3)
F-BHBR	1649A-98-11	1036	Air France
F-BHBS, BHBT	1649A-98-11	1044, 1045	Air France**
F-BHMI/BHML	1049G-82	4668/4671	Air France (4)
F-BRAD (4)	1049G-82	4519	From C-55-81 via E-55-01
F-BRNH (2)	1049	4513	From C-55-81 via E-55-01
F-ZVMV (5)	749-79-22	2503	French Air Force

Registration	Model	C/N	Remarks
Great Britain (G)			
G-AHEJ/AHEM	C-69-5/049E-46	1975/1978	42-94554/94557 to BOAC (4)**
G-AHEN (1, 4)	C-69-5/049	1980	42-94559 to BOAC; later as 049D**
G-AKCE (2)	C-69C-1/049E-46	1971	Ex42-94550
G-ALAK, ALAL (2)	749A-79	2548, 2549	From 749-79-32
G-ALAM,ALAN (2)	749A-79	2554, 2555	From 749-79-32
G-AMUP (2)	049-46-27	2051	
G-AMUR (2)	049E	2065	From 049-46-27
G-ANNT (2)	749A-79-52	2671	
G-ANTF, ANTG (3)	749A-79	2504, 2505	From 749-79-22
G-ANUP (2)	749A-79	2562	From 749-79-31
G-ANUR (2)	749A-792565		From 749-79-31
G-ANUV (3)	749A-792551		From 749-79-33
G-ANUX, ANUY (3)	749A-79	2556, 2557	From 749-79-33
G-ANUZ (3)	749A-79	2559	From 749-79-33
G-ANVA (3)	749A-79	2564	From 749-79-33
G-ANVB (3)	749A-79	2589	From 749-79-33
G-ANVD (3)	749A-79	2544	From 749-79-33
G-ARHJ (2)	049-46-27	2051	
G-ARHK (5)	049-46-26	2036	
G-ARVP (5)	C-69-1/049	1967	Ex 43-10315
G-ARXE (6)	C-69-1/049	1965	Ex 43-10313
G-ASYF (2)	749A-79-50	2630	
G-ASYS (2)	749A-79-50	2623	As freighter
G-ASYT, ASYU (2)	749A-79-50	2631, 2632	

* Leased to Air Afrique as TU-TBB.
** Last Constellations built.

Registration	Model	C/N	Remarks
Switzerland (HB)			
HB-IBI			
HB-IBJ			
HB-IEA	C-69-1/049	1965	Ex 43-10313
HB-IEB	C-69-1/049	1967	Ex 43-10305
HB-IED	C-69-1/049	1980	Ex 42-94559
Haiti (HH)			
HH-ABA (2)	749-79-12	2615	
Dominican Republic (HI)			
1-129 (3)	649-79-12		
1-140 (2)	649-79-12		
1-207 (4)	649-79-12		
1-228 (2)	1049-53-67		
1-254 (2)	1049H		
1-26 0 (3)	049-46-59		
1-27 0 (2)	049-46	2085	
HI-328 (2)	C-121A 2	607	Ex 48-614
HI-329 (2)	1049C-55-83	4536	Sprayer
HI-332 (4)	649-79-12	2522	

Registration	Model	C/N	Remarks
Colombia (HK)			
HK-162, 163	749A-79-74	2663, 2664	Avianca
HK-175/177	1049E-55	4554/4556	Avianca
HK-184	1049G-82	4628	Avianca
HK-650 (4)	749A-79	2544	From 749-79-33
HK-651 (4)	749A-79	2557	From 749-79-33
HK-652 (4)	749A-79	2564	From 749-79-33
HK-653 (2)	749A-79-52	2645	
South Korea (HL)			
HL-102 (5)	749A-79	2551	From 749-79-33
HL-4002 (2)	1049H	4819	
HL-4003 (2)	1649A-98	1037	Freighter
HL-4006 (2)	1049H	4816	
Panama (HP)			
HP-280, 281	1049G-82	4677, 4678	Thai Airwasy (2)
HP-467 (7)	1049G-82	4678	
HP-475 (3)	1049E-55	4551	
HP-501 (5)	1649A-98-11	1036	
HP-526 (2)	1049H	4815	
Thailand (SVM, HS)			
HS-TCA (#1)	1049G-82	4644	Thai Airways
HS-TCA (#2)	1049G-82	4672	Thai Airways
HS-TCB (#1)	1049G-82	4645	Thai Airways
HS-TCB (#2)	1049G-82	4677	Thai Airways
HS-TCC	1049G-82	4678	Thai Airways
Argentina (LV)			
LV-FTU, FTV	1049H	4846, 4847	Transcontinental SA (Lse.)
LV-IGS (4)	749-79-33	2540	
V-1IC (5)	749A-79-44	2619	
LV-ILN (2)	1649A-98	1030	
V-ILW (3)	1049D	4166	
V-IXZ (5)	1049G-82	4580	From 1049G-55-01
LV-JHF (4)	1049H-82	4801	
LV-JIO (5)	1049H-82	4808	
LV-JJO (5)	1049H-82	4807	
V-PBH (4)	749A-79-44	2619	Freighter
V-PCQ (2)	1049D	4166	
LV-PJU (3)	1049H-82	4801	
LV-PKW #1 (3)	049-46-59	2069	
LV-PKW #2 (4)	1049H-82	4807	
LV-PZX (3)	749-79-33	2540	Freighter

Registration	Model	C/N	Remarks
Luxembourg (LX)			
LX-IOK (4)	749-79-31	2562	
LX-LGX (4)	1649A-98	1042	
LX-LGY (2)	1649A-98-11	1036	
LX-LGZ (4)	1649A-98	1041	
United States (NC, N)			
N65 (2)	749A-79-52	2648	
N119 (1st civ)	PO -1W/WV-1	2612	Ex Navy 124437 to FAA
N120 (1st civ)	749A-79-43	2613	Ex Navy 124438 to FAA
N121 (2)	749A-79-52	2654	
N964 (5)	1049G-82	4683	
N1102 (2)	1649A	1001	
N1192 (2nd civ)	PO -1W/WV-1	2612	N119 Renumbered for USAF
N1206 (2nd civ)	PO-1W/WV-1	2613	N120 Renumbered for USAF
N1649	1649A	1001	Lockheed test
N1880	1049H	4820	Dollar Associates (lease)
N1949 (4)	749-79-31	2565	
N4624 (2)	1049G-82	4617	
N4796 (3)	1649A	1036	
N8021 (2)	1049G-82	4673	
N8022 (2)	1049G-82	4676	
N8023 (2)	1049E-55	4551	
N8024, 8025 (4)	1049G-82	4644, 4645	
N8026 (6)	1049G-82	4678	
N8338 (2)	1049G-82	4616	
N9463 (2)	VC-121A	2602	Ex 48-610
N9464 (1st civ)	C-121A	2601	Ex 48-609
N9465 (1st civ)	C-121A	2604	Ex 48-612
N9466 (1st civ)	C-121A	2607	Ex 48-616
N9467 (1st civ)	C-121A	2609	Ex 48-617
N10401 (3)	749A-79-33	2661	
N10403	749A-79-33	2622	
N25600	C-69/049-39-1O	1961	1st Constellation
N38936	C-69-1/049	1962	Ex 43-10310, as 049
N45511 (2)	1649A-98	1034	
N45512 (2)	1649A-98	1040	
N45515 (2)	1049H	4843	
N45516 (2)	1049H	4840	
N45517 (2)	1649A-98	1041	
N45520 (2)	1649A-98	1042	
N51517 (3)	1049G-82	4618	

Registration	Model	C/N	Remarks
United States (NC, N)			
N54212	C-69C-1/049	1971	Ex 42-94550, as 049
N54214	C-69-5/049	1974	
N67900	C-69/049	1961	
N67952	C-69-1/049	1963	
N67953	C-69-1/049	1964	
N70000	C-69-5/049	1974	
N73544	C-121C	4175	Ex 54-156
N86500, 86501	C-69-1O	2021, 2022	Ex 42-94560,
N86502/86509	049-46-25	2023/2030	TWA (8)
N86510, 86511	049-46-26	2034, 2035	TWA
N86512/86517	049-46-26	2039/2044	TWA (6)
N86520	749-79-22	2503	Lockeed to Aerovias
N86521	649A-79-60	2642	Chicago & Southern
N86522	649A-79-60	2653	Chicago & Southern
N86523, 86524	649A-79-60	2659, 2650	Chicago & Southern
N86525	649A-79-60	2662	Chicago & Southern
N86526	049-46	2084	Lockheed to KLM
N86527/86530	749-79-22	2525/2528	Pan Am (4)
N86531, 86532 (2)	049-46-59	2068, 2069	
N86533 (3)	049-46-59	2071	
N86535	649A-79-60	2673	Chicago & Southern
N86536	C-69-5	1979	Ex 42-94558, as 049
N86682	1049G-82-118	4679	Qantas
N88831/88833	049-46-26	2031/2033	Pan Am (3)
N88836/88838	049-46-26	2036/2038	Pan Am (3)
N88845/88850	049-46-26	2045/2050	Pan Am (6)
N88855/88862	049-46-26	2055/2062	Pan Am (8)
N88865	049-46·26	2066	Pan Am
N88868	049-46-26	2067	Pan Am
N90607 (2)	749-79-33	2551	
N90608 (2)	749-79-33	2564	
N90621 (2)	749-79-33	2559	
N90622 (2)	749-79-33	2544	
N90623 (2)	749-79-33	2556	
N90624 (2)	749-79-33	2589	
N90625 (2)	749-79-33	2557	
N90814/90826	049-46	2076/2088	TWA (13)
N90827	C-69-1	1965	Ex 43-10313, as 049
N90828/90831	C-69-1	1967/1970	Ex 43-10315, 10316, as 049
N90921/90924 ·	049-46-27	2051/2054	American Overseas (4)
N90925/90927	049-46-27	2063/2065	American Overseas (3)
N91201/91212	749-79-22	2577/2588	TWA (12)
N93164	1049G-82	4581	Freighter, from 1049E-55-01

Registration	Model	C/N	Remarks
United States (A, AS and AV)			
N101/N107A	649-79-12	2518/2524	Eastern (7)
N108/N110A	649-79-12	2529/2531	Eastern (3)
N112/N114A	649-79-12	2533/2535	Eastern (3)
N115A, 116A	749-79-1,2	2610, 2611	Eastern (2)
N117/N121A	749-79-12	2614/2618	Eastern (5)
N2717A (2)	749A-79-46	2513	From 749-79-22
N2727A (2)	C-69-5	1976	Ex 42-94555, as 049E-36
N2735A (2)	C-69-5	1978	Ex 42-94557, as 049E-46
N2736A (2)	C-69-5	1977	Ex 42-94556, as 049E-46
N2738A	049-46-27	2051	American Overseas
N2739A	049-46-27	2065	American Overseas
N2740A(2)	C-69-5	1975	Ex 42-94554, as 049E-46
N2741A (3)	C-69C-1	1971	Ex 42-94550, as 049E-46
N4192A (3)	1049G-82	4581	From 1049E-55-01
N5595A (3)	749A-79-44	2620	
N5596A (3)	749A-79-44	2619	As freighter
N608AS	C-1218	2600	Ex 48-608, as sprayer
N179AV (3)	1649A-98	1040	
United States (B, C and CS)			
N2520B, 2521B (3)	049-46	2081, 2082	Ex 90819, 90820, YV-C-AME, AMI
N468C, 469C (2)	1049H	4846, 4847	
N661C (2)	1049G-82	4602	
N1005C (2)	1049E-55	4557	
N1006C	1049H-82	4802	Seaboard & Western
N1007C/1010C 1	049H-82	4805/4808	Seaboard & Western (4)
N4900C (2)	749A-79-74	2663	
N4901C (3)	749A-79-52	2671	
N4902C (3)	749A-79	2566	From 749-79-32
N4903C	1049G-82	4619	Howard Hughes
N6000C (2)	049-46-59	2020	
N6001C/6005C	749A-79-52	2633/2637	TWA (15)
N6006C	749A-79-52	2639	TWA
N6007C/6015C	749A-79-52	2643/2651	TWA (9)
N6016C/6020C	749A-79-52	2654/2658	TWA (5)
N6021C/6026C	749A-79-52	2667/2672	TWA (6) 6025C to Howard Hughes
N6215C/6218C	1049C-55-83	4523/4526	Eastern (4)
N6219C/6230C	1049C-55-83	4527/4538	Eastern (12)
N6501C/6504C	1049D	4163/4166	Seaboard & Western (4)

Registration	Model	C/N	Remarks
United States (B, C and CS)			
N6695C (4)	749A-79-82	2671	
N6696C (2)	749A-79-52	4619	
N6901C/6910C	1049-54-80	4015/4024	TWA
N6911C	1049H	4804	Flying Tigers
N6912C/6915C	1049H	4809/4812	Flying Tigers (4)
N6916C/6918C	1049H	4814/4816	Flying Tigers (3)
N6919C	1049H	4819	Flying Tigers
N6920C	1049H	4822	Flying Tigers
N6921C	1049H	4817	Air Finance Corp.
N6922C	1049H	4825	Air Finance Corp.
N6923C	1049H	4826	California Eastern
N6924C, 6925C	1049H	4852, 4853	Flying Tigers
N6931C	1049H	4813	California Eastern
N6932C	1049H	4823	California Eastern
N6933C	1049H	4826	California Eastern
N6935C, 6936C	1049H	4848, 4849	Slick Airways
N6937C	1049H	4830	Slick Airways
N7023C (3)	1049H	4818	
N7101C/N7120C	1049G	4582/4601	TWA (20)
N7121C/7125C	1049G-82	4648/4652	TWA (5)
N7301C/7309C	1649A-98	1002/1010	TWA (9)
N731OC /7317C	1649A-98	1012/1019	TWA (8)
N7318C/7322C	1649A-98	1021/1025	TWA (5)
N7323C, 7324C	1649A-98	1029, 1030	TWA
N7325C	1649A-98	1035	TWA
N7772C (3)	1049G	4682	
N7776C (2)	1049H-82	4801	
N7777C (2)	1049H-82-133	4803	
N9714C (2)	1049-55-01	4580	
N9716C, 9717C (2)	1049E-55-01	4545, 4546	From 1049C-55-81
N9718C (2)	1049C-55-01	4549	From 55-81
N9719C (2)	1049E-55-01	4574	
N9720C (2)	1049G	4578	From 1049E
N9721C (2)	1049G	4607	
N9722C (2)	1049G-82-118	4679	
N9723C (2)	1049G-82-118	4680	
N9751C (3)	1049G-82	4607	
N9752C (2)	1049H	4850	
N22CS (5)	749A-79	2556	From 749-79-33
United States (D, E, F and G)			
N833D (5)	1049G-82	4677	
N563E, 564E (2)	1049H	4833, 4834	
N565E, 566E (2)	1049H	4837, 4838	
N9970E	VC-121A	2602	Ex 48-610
N9812F (3)	749A-79	2559	From 749-79-33
N9813F (4)	749A-79	2589	From 749-79-33

Registration	Model	C/N	Remarks
United States (D, E, F and G)			
N9816F (4)	749A	2504	From 749-79-22
N9830F	749A-79	2551	From 749-79-33
N4715G (2)	1049C-55	4543	
US Army/Air Force Constellations			
42-94549/94559 (11)	C-C9-5 1	970/1980	
42-94560, 94561 (2)	C-C9-5	2021, 2022	
43-10309 (1)	C-69	1961	
43-10310/10317 (8)	C-C9-1	1962/1969	
48-608 (1)	VC-121B	2600	
48-609/617 (9)	C-121A	2601/2609	
51-3836/3845 (10)	RC-121C	4112/4121	
52-3411/3425 (15)	RC-121D	4329/4343	
53-533/556 (24)	RC-121D	4348/4371	
53-3398/3403 (6)	RC-121D	4372/4377	
53-78851	VC-121E	4151	Ex Navy R7V- 1131650
53-8157, 8158 (2)	YC-121F	4161, 4162	Ex-Navy RTV-2
54-151/183 (33)	C-121C	4170/4202	
54-2304/2308 (5)	RC-121D	4386/4390	To EC-121D in 1962
55-118/139 (22)	RC-1210	4391/4412	To EC-121D in 1962
54-4048/4079 (32)	C-121G		Ex-Navy R7V-1
67-21471,214722	EC-121R	4382, 4385	Ex-Navy EC-121P
67-21473/21500 (28)	EC-121R		Ex-Navy EC-121K
US Navy Constellations			
124437, 124438 (2)	P0-1W	2612, 2613	
126512, 126513 (2)	WV-2	4301, 4302	
128323/128326 (4)	WV-2	4303/4306	
128434/128444 (11)	R7V-1	4101/4111	
131387/131392 (6)	WV-2	4307/4312	
131621/131629 (9)	R7V-1	4122/4130	
131632/131659 (28)	R7V-1	4133/4160	
135746/135761 (16)	WV-2	4313/4328	
137887/137890 (4)	WV-2	4344/4347	
137891/137898 (8)	WV-3	4378/4385	
140311/140313 (3)	R7V-1	4167/4169	
141289/141333 (45)	WV-2	4413/4457	
143184/143225 (42)	WV-2	4458/4499	
143226/143230 (5)	WV-2	5500/5504	

SR Technics

20 Bibliography

Breffort, Dominique and André Joiuneau, *Lockheed Constellation (Sky Legends)*, Histoire & Collections, 2005.

Brown, Austin J, *Lockheed Constellation*, Ian Allan, 1993.

Cohen, Lizabeth, *A Consumer's Republic: The Politics of Mass Consumption in Post-war America,* Vintage Books, 2004.

Corrie, Nick, *The Connie Breed — Lockheed Constellation,* Galago, 1993.

Germain, Scott E, Lockheed *Constellation and Super Constellation: 1 (AirlinerTech S.),* Specialty Press, 1998

Henderson, Scott, *Lockheed Constellation in Colour: A Photographic History of One of the Most Charismatic American Civil Aircraft Ever Built,* Scoval Publishing, 2007.

Marson, Peter J, *Lockheed Constellation*, Air Britain, 1983.

Peters, P M, J. Middel and A. Hoolhorst, *Fuel Efficiency of Commercial Aircraft: An Overview of Historical and Future Trends,* National Space Laboratory, 2005.

Serling, Robert J, *The President's Plane is Missing,* Cassell & Co., 1968.

Serling, Robert J, *Howard Hughes' Airline: An Informal History of TWA,* Lume Books, 1983 [2018].

Stringfellow, Curtis K and Peter M Bowers, *Lockheed Constellation: Design, Development, and Service History of All Civil and Military Constellations, Super Constellations, and Starliners,* Motorbooks International, 1992.

Winchester, Jim, *Lockheed Constellation* (Airlife's Classic Airliners S.), Crowood Press, 2001.

Wixey, Kenneth E, *Lockheed Constellation* (Classic Civil Aircraft S.), Ian Allan, 2007.

VH EAB
QANTAS
super service at your command !
QANTAS
Australia's International Airline
QANTAS EMPIRE AIRWAYS LTD. (Incorp. in Qld.) IN ASSOCIATION WITH BRITISH OVERSEAS AIRWAYS CORPORATION AND TASMAN EMPIRE AIRWAYS LTD.

INDEX

MOST COMFORTABLE WAY TO GET THERE FAST

QUIET LUXURY TO MAKE THE TIME FLY—NEW SPEED TO SHORTEN THE DISTANCE

Largest, Roomiest Airliner in the World   Far Quieter for Greater Comfort   Wider Aisles and Seats

Larger Windows   Finest Air Conditioning   Restful 5-Cabin Privacy

Congenial Starlight Lounge   Henry Dreyfuss Interiors   Fastest Constellation Ever Built

For all the speed, and quiet comfort, too, fly Super Constellations over every ocean and continent on these 20 leading airlines: AIR FRANCE · AIR-INDIA INTERNATIONAL · AVIANCA · CUBANA · DEUTSCHE LUFTHANSA · EASTERN AIR LINES · FLYING TIGER LINE · IBERIA · KLM · LAV · NORTHWEST ORIENT AIRLINES · PAKISTAN INTERNATIONAL · QANTAS · SEABOARD & WESTERN · SLICK AIRWAYS · TAP · THAI AIRWAYS · TRANS-CANADA AIR LINES · TWA-TRANS WORLD AIRLINES · VARIG

LOCKHEED SUPER CONSTELLATION

Look to Lockheed for Leadership

PP-VDE
VA

VARIG

LU